W9-CIN-572

The Emergence of Life

THE EMERGENCE OF LIFE

Darwinian Evolution from the Inside

SIDNEY FOX

Basic Books, Inc., Publishers

NEW YORK

Library of Congress Cataloging-in-Publication Data

Fox, Sidney W.
The emergence of life.

Includes index.
1. Life—Origin. 2. Evolution. I. Title.
QH325.F658 1988 577 87–47778
ISBN 0–465–01925–0

Printed in the United States of America
Designed by Vincent Torre
88 89 90 91 HC 9 8 7 6 5 4 3 2 1

CONTENTS

PREFACE

As an author, I ask myself, Who is the audience for this book? My experiences tell me that the subject I know best has no single, homogeneous readership. For discussions of how life began, there is only a mosaic of audiences, each with its own interests and perspectives. Authors can hardly write on this subject for a single collective group. Perhaps the best that one can do is to write for one set of readers mainly and attempt to see that others will find the treatment sufficiently tasteful to derive something of value.

The primary audience I have selected is that composed of those who have had a high school course in biology. Thanks to numerous well-oriented authors of textbooks for that level, large numbers of students have come partially to recognize that life and its beginning are natural processes, and not miracles in the strict sense of that word.[1]

For someone who knows little of evolution, the visible result of the evolutionary sequence is often considered equivalent to miraculous. Accordingly, this book seeks to answer primarily the kind of question that younger and nontechnical audiences have asked. It attempts to explain evolutionary theory to the uninitiated, and scientific methodology to those who will not become vocational scientists. It attempts to clarify the essential nature of the transition from the inanimate to the animate, and it relates personal experiences of the kind with which intelligent readers can identify.

The advances and perceptions reviewed in this book are not an impartial selection of various points of view on how life emerged on this planet. They are instead a treatment of the connected picture the author sees as most relevant—the one most emphasized in our laboratory relative to observations by many others in the field. Lack of subjectivity cannot be claimed. In striving to develop an accurate picture for myself, I have conscientiously employed and questioned those criteria of accuracy that I was introduced to early in my fact-and-document, chemistry-notebook style of education.

Where I differ markedly from others who approach this problem is in the choice of the method of investigation, of research and interpretation. This mode was developed essentially as an attempt to retrace the pathway of early evolution from the beginning chemical substances through the construction of cells. Others attempt to extrapolate backward from the present situation analytically or to reason from the quite sure information of the present world.

My feeling is also much like the one Steven Weinberg expressed in his book *The First Three Minutes,* in which he says, "I do not believe that scientific progress is always best advanced by keeping an altogether open mind. . . . The great thing is not to be free of theoretical prejudices but to have the right theoretical prejudices. And always, the test of any theoretical preconception is in where it leads."[2]

I feel comfortable with a frankly acknowledged bias derived from experiments and experience. That is because it appears to me that the universe and life in it began with a bias; it is not true that all (now) conceivable alternatives had an equal opportunity. To the extent that some possibilities were more pregnant for development than others, the universe originated and developed from bias. At least in this way, I feel in tune with my universe and with, I hope, its special biases.[3] While Weinberg cannot experiment with galaxies and planets, I as a chemist can experiment with proteins, nucleotides, and cells. Those of us who experiment with a retracement of aspects of the emergence of life find that the experiments themselves often contribute richly to identifying new processes, new phenomena, and a new philosophy.

The book and the author are indebted to the following for their help in ensuring technical accuracy, for general clarity, and, in several instances, for the development of the state of the art also. These include Steven Brooke, Robert G. Cuff, Jr., Klaus Dose, J. L. Fox, Kaoru Harada, P. E. Hare, Franz Hefti, Mae-Wan Ho, John Jungck, Anwar Khoury, James C. Lacey, Jr., Koichiro Matsuno, Mavis Middlebrook, Tadayoshi Nakashima, Robert Shapiro, Graham Vaughan, Allen Vegotsky, and Richard S. Young. Of course, the responsibility for what is finally stated in the book is totally that of the author.

My special gratitude is expressed to Duane L. Rohlfing, Koichiro Matsuno, and Larry Fox for their care in technical reviews of the total book. Professor Rohlfing's reminders that communicating with a nontechnical audience carries added responsibility and his having "been there" in the early days of testing alternatives are deeply rooted memories. Professor Matsuno deserves full acknowledgment for his wide-angle understanding and his view from the physicist's roost. The editorial efforts of Richard Liebmann-Smith of Basic Books have also given me thought-provoking pleasure. To travel with all of the above people, and with many others,* in various ways has been a life-filling joy.

*Of those patrons who have supported the research by providing equipment and compensation to those of the above who have labored in the laboratories, my colleagues and I are outstandingly indebted to the National Aeronautics and Space Administration for Grant NGR 10-007-008. Other individuals not listed above, and whose work has contributed to the results described, are acknowledged in scientific publications. I wish to express my warmest thanks to the erstwhile president of the University of Miami. Dr. Henry King Stanford understood the research as it was evolving, he understood the need to protect it, and he protected it.

The Emergence of Life

Conversation 1

The questions I most often meet in an informal situation are the following:

What do you do?

I'm a biochemist.

What kind of biochemistry do you work on?

I work in the problem area of the origin of life and its early evolution.

How is that coming?

Rather well, but the perception by others is very uneven. It runs all the way from the notion that there's nothing yet to replace miraculous origins to the conviction that the problem has been solved. At one extreme, a number of Ph.D.s believe it's folly to try to solve it; some still believe it's sacrilegious to try.

What do you think?

The most important first step was to move the problem from its traditional context of a miracle to that of holistic, stepwise, evolutionary science. That has to be done totally, including subliminally, by the scientist; the question is then *rephrased* as many questions in a stepwise sequence.

Do you know the sequence?

In a sense, I emphasize that we didn't come to know the sequence. To the degree that the essence of the sequence problem is solved, it essentially solved itself—years before we realized what had already happened in the laboratory.

We do know the key steps in the laboratory retracement of inanimate matter to organized cells; others can and do repeat those steps at will. For me, a full solution will be to retrace inanimate matter to man and to human society in extensive and exquisite detail. But yes, I think the key steps in the transformation from inanimate matter to organized cells are understood in considerable detail. The principles of evolution to human beings, to abstract thought, and to society can now be more clearly perceived because the initial steps are as well charted as they are.

What are the main questions and answers about the first steps?

In a broad sense, the first steps revealed by experiment describe how a really complex form of inanimate matter, proteinoid (thermal protein), arose quickly and simply from simple precursors, and how that matter was converted to organized microspheres.

What are those?

Proteinoid is a complex type of protein-like matter, and microspheres are the cell-like units that form when water comes into contact with proteinoid. They fill the role of the most primitive type of cell having properties of life. The unit of life is the cell; the unit of protolife (first life) is the protocell.

Why have these questions posed so much difficulty in the past?

The most important question to grasp was how to work on the problem. The answer to that was itself an evolution. The only way yet known is to retrace the steps from inanimate matter to organized cells and beyond. Analyzing what's here now can't replace experimenting in the direction of evolution itself—building up from the right beginning molecules in a narrow range of choices. Choices *by* molecules preceded life and extended into life, where chemistry is still fundamental. As evolution continued from molecules alone in the first stage, there was some fine-tuning in the large molecules of protein and nucleic acids, and some "trial-and-error" events occurred as well. Realization of all that has taken time and much integrated overview in research. The synthetic approach of evolution is not yet a significant principle in our educational system.

Conversation 1

When are you going to make something that crawls out of the test tube?

That question has often been asked by both scientists and journalists. I think of it as subliminally an extrapolation of thinking from the age of miracles; it requires essentially an act of instant creation. The question itself has to be rephrased in the modern evolutionary context, in which development is step by step, not instant and not gradual, but sudden step by sudden step. For each step, the right components have to come together in the right way. The most critical question is, or was, When does inanimate matter get converted in the series of step-by-step changes into something that can be said to have living qualities?

Why do you sometimes talk about making a few amino acids as if they were the secret of life?

No one in our group has ever done that. I agree with you that such a jump is an unwarranted extrapolation. Even though amino acids, as the components of protein, are essential to life, very much more about them needs to be understood. We now know which amino acids are necessary for the most useful stable cellular structure, which ones for one kind of enzyme action, which ones for another kind, which ones for selective membrane activity, and so on. We also know that certain conditions were essential step by step for the very involved sequence of transformations to living systems. For many fringe questions, we would like better answers, but the central queries have been answered by the identification of internal—that is, true evolutionary—processes. If any one of these processes is the "secret" of life, I would say it's the high degree of self-ordering of amino acids, and the resultant diversity of functions.

How long has research on the origin of life been going on?

For centuries. In the last century, the research was on what was known as spontaneous generation. A number of scientists then and later had what have come to be recognized as very appropriate ideas, along with confusing ones in some cases. But many deserve to be credited.[1]

CHAPTER 1

Back to the Future

Where did I come from? How am I here? Why am I me? Such are the questions we ask from an early age.

As we grow older, our questions become more sophisticated: What, rather than where, do I come from? What will I become? What is my relationship to others? The later questions are more materialistic, more outgoing, more socially oriented.

As we gain perspective, the questions tend to move from I to we, then to we humans, we animals, and we living things. The perspective of the essential questions narrows from the outside to the inside of organisms, and it goes back in time. We become more analytical. While development of analytical ability is normal, it runs counter to the direction of the answers to the questions, which must be composed in a synthetic, forward direction. The question that moves from us as individuals to humans to animals to living things is analytical, and it is also a kind of extrapolation backward for which we learn to position our mind where we can look forward to the present.

The answers are to be found along a climbing evolutionary tree from simple to complex. The curiosity that we share has existed throughout recorded history. Efforts to answer questions of genesis became laboratory efforts in the nineteenth century, with experiments on spontaneous generation, one of the activities with which the name of Louis Pasteur is associated.

The modern era of scientific investigation, whose first results attained wide educational dissemination, can be said to have begun in the early 1960s. In 1963, the U.S. Biological Sciences Curriculum Study (BSCS) first published a teaching book, *Molecules to Man,* for use in high school biology courses.[1] The production of this book, under the direction of Arnold Grobman, involved over one hundred experts in biological education who met for many years in Colorado. It has since been adopted or copied in numerous foreign countries.

The BSCS book was a first textbook to include laboratory representations of what might have been the first cells on Earth, as seen through the ordinary microscope. The book also showed photographs through the microscope of an earlier representation of first cells, known as coacervate droplets. The latter were used by the Soviet biochemist Alexander I. Oparin in representations of the first *pre*living cells. In addition, this new educational material discussed one way in which the simplest building blocks of protein (which are in turn the principal building blocks of cells) might have arisen on the primitive Earth.

Advances in science are often forecast by earlier, less complete, developments that are rediscovered later. This area is no exception. Alfonso L. Herrera, a Mexican cosmologist, reviewed his experiments in 1942 in the U.S. journal *Science.* He had produced amino acids, protein-like polymers, pigments, and loose clusters of molecules which bore much resemblance to cells.

Polymer molecules are simply large molecules built repetitively from small ones, monomers. Proteins are polymers composed of small molecules, amino acids. We now know that Herrera's polymers could have been primitive kinds of protein. The 1963 BSCS book appeared, therefore, after a preliminary but significant kind of advance. Herrera's pioneering was not widely appreciated at the time, however. It is still undervalued, even by Mexican scientists.

Mythology and Legend

The history of explanations for the existence of life on this planet necessarily begins with the mythological stage. The most primitive kind of thinking is the mythological, represented in legends and holy books. At the other extreme is the set of inferences from scientific detective work. At the same time that the scientific work proceeds, the mythological explanations hold the majority of listeners on this planet. To a large extent, as a result, the transitional era in which we live is not calm. The general thinking is caught in crosscurrents. Of such stuff are intellectual storms composed.

Channeled Thinking in Science

The mechanism of everyday thought has much in common with Darwinian selection. In each process, many variants are generated, and one or a few are selected, or chosen. The variants in thoughts are ideas; the variants in organisms are mutants. If the generation of ideas or of mutants is large, the selection process has more to choose from. The greater the variation, the richer the thinking—or the richer the potential for evolution.

At this point, we may profitably introduce a clarifying emphasis made by Darwin—an emphasis many of his disciples have lost sight of. Having derived much of his thinking from artificial selection, that is, from breeding practices, Darwin chose the term *natural selection* to indicate those instances in which the favored variant was "selected" by the environment instead of by the purposive human breeder. At the same time, he recognized that the environment was acting passively rather than actively, that is, was not acting toward a goal. He stated, in fact, that a better term than *selection* was *preservation.* In later editions of *The Origin of Species,* after he saw how his words were being misinterpreted, he used the term *Unconscious Selection,* placed *natural selection* in quotation marks, and also tended to stress *preservation* rather than selection. For example, he said, "Several writers have misapprehended or objected to the term Natural Selection. Some have even imagined that natural selection induces variability whereas it implies only the preservation of such variations as arise and are beneficial to the being under its conditions of life."[2]

These later emphases of Darwin's have proved to agree with experiments in the middle of the twentieth century. The study of evolution at the level of molecules began more recently; in part, it is what I mean by evolution *from the inside.* Our concern here is with a level lower than that of the biological variations that happen to be best adapted to the circumstances suitable for life in the environment. We focus on varieties of molecules that are best adapted to their immediate environment, which is other molecules. In the myriad molecular interactions that are possible, because of the tremendous variety of shapes that can be theoretically visualized for molecules, only a very minor fraction is chosen and endures. This process is Molecular Selection, which will be discussed in detail later, especially in chapter 8. Insofar as human consciousness is concerned, molecular selection is unconscious selection, much as Darwin defined Natural Selection.[3] Molecular Selection, however, is active, in contrast to Natural Selection. That is to say, choices between molecules are manifest in their actions. In natural selection, choices between organisms are due to the passive sifting action by the environment. Both molecular selection and natural selection are natural, strictly speaking. But the results of the biological type are directly visible to the eye, whereas those of the molecular type are not.

What emerges with clarity from studies of molecules is that once the "selection" is made, the highway on which it proceeds is a narrowed one. New molecular individuals are very much limited in type by boundaries set at earlier stages.

Relevant to everyday thought is the reality that our thinking is greatly affected by traditional intellectual boundaries, even though we may believe on first observation that our reasoning is uninfluenced by them.

When these processes are looked at from the perspective of Unconscious Selection and when a modern organism is traced from its evolutionary beginning or a thought from its origins, they are seen to be channeled. The effect of channeled thinking in science and elsewhere is more profound than is ordinarily realized.

My own first, vivid introduction to an awareness of channeled thinking occurred when T. H. Morgan and I were discussing a choice of fields for my graduate study. Morgan was the chairman of my Ph.D. committee. He was famous for the chromosomal theory of the gene, for which he was awarded in 1933 the first Nobel Prize given in genetics. Morgan has been recognized as a zealous recruiter of physicists and chemists into biology. In our discussion, I raised doubts about doing graduate work in biology instead of chemistry, partly since I had taken no undergraduate course in biology. Morgan's response was that biological research needed people who were trained in the physical sciences and whose thinking had not been channeled in the old biological ruts. Morgan was very sensitive to the power of channeled thinking.

Incidentally, such formal questions as where to do graduate study were usually taken up under informal conditions—during sessions on weekends in which my wife and I accompanied Morgan and a former student of his, Professor Albert Tyler, to Caltech's Corona del Mar marine laboratory. Morgan would set up an experiment with, typically, 64 egg dishes to study fertilization in sea squirts, then expand it stepwise to 128 by the new tests and controls he visualized as he assembled dishes. Since sea squirts fertilize themselves, Morgan was interested in removing the block to self-fertilization. He reasoned that this had to occur chemically in nature, and that was a main reason for my association with these experiments. Discussions of the related questions were mixed in with such matters as the decision on graduate study, and while we all ate lunch standing up. Under such conditions, it was necessary to compartmentalize thinking!

Finding a solution to the problem of life's origin has been the victim of prior channeled thinking, much more widely than have other scientific questions. This kind of influence therefore deserves some attention here. Channeled thinking occurs in all areas of activity, normally with no intent to preempt our perceptions, although sometimes that is the objective. Often there is no ulterior intent at all; much of the molding of perceptions is through subliminal processing. Channeled, or preemptive, thinking is not limited to politics, in which it is anticipated. That formidable stateswoman Margaret Thatcher spoke of this phenomenon a few years ago when she observed, "I do sometimes think that words are used in such a way that they influence people's perceptions and because of words they do not look at what there is to see—they look and see what propaganda has taught them to look for and find."[4] *Propaganda* may seem like a harsh word for teaching a favored view, especially in science, but the dividing lines between education, indoctrination, conditioning, and propaganda are not sharp. The channeling principle applies throughout, and this is what the scientist Morgan recognized and the stateswoman Thatcher commented on in their own fields.

Where Did Amino Acids Come From?

An example of preemption occurred in our area of research when information was released to the public on the findings from the Apollo program. Our part of the Apollo program was to analyze samples returned by the astronauts from the Moon for amino acids, the building blocks of protein molecules. The question, in essence, was how far prebiotic (preliving) evolution had proceeded along the chemical trail to life. For instance, had it attained the amino acid stage, or had it come to be initially recognizably related to life by yielding protein molecules?

Although the question "Where did we come from?" is probably the most popular way of phrasing the central issue of mankind's interest in its own origin, we have learned that we must rephrase that question into chemical terms: *What* did we come from? As we answer the chemical question of *what,* we are better able to deal with the question of *where.* The where question, however, then becomes, Where could, or did, those needed chemicals come from?

To answer the question of what we, as late evolutionary representatives of life, came from, we can examine what living things are made of, and we can also catalog what they need to eat in order to survive and reproduce. This basic information is the surest we can employ in trying to answer our question. One might fantasize that the original living things were of a chemical nature different from that of today's living things, and a few theorists have, in fact, argued just that by suggesting that the precursors of modern life were clay organisms or were composed of organic silicon compounds. However, no evidence has been found to suggest how inorganic clay was replaced by the organic material common to all living things known today. Perhaps more significant is the criticism that there are essentially no vestiges of clay or organic silicon, as such, in life as we know it, despite some vestiges of other specific minerals.

When we attempt to infer the composition of original organisms, our surest approach is to ask what all extant organisms are composed of and what they must eat. These questions have been posed and answered extensively in this century, especially between 1920 and 1965. The answers have been compressed into a single phrase, *the unity of biochemistry.* [5] What all living things are composed of, what they need, and what they must eat is remarkably unified, and also unifying for our understanding.

In short, all living things require the same materials, obtained either in the diet or by internal manufacture. First and foremost is water. Water is the universal solvent within each cell, and all active living organisms require replenishment of water.

Second in amount on our list is protein. Plants are an exception; they are composed primarily of cellulose and other carbohydrates. Most individuals are familiar with the need for protein and with the use of carbohydrate for energy. The general concept of nutrition stresses the need to replenish protein, obtained from meat, fish, eggs, milk, grains, beans, nuts, and vegetables. Those proteins were all once organisms or derived from organisms, and the lower organisms were themselves composed predominantly of water and proteins.

In addition to water and protein, the stuff of life includes carbohydrates, nucleic acid (DNA/RNA) required for inheritance, vitamins, lipids (fats), and minerals.

These nutritional requirements and these compositions apply to all evolutionary levels of living things; hence *the unity of biochemistry.* This unity was already recognized by Charles Darwin and was resuggested several times by biochemists as knowledge grew. In the 1960s, it was massively documented, especially by Marcel Florkin, a Belgian biochemist. Florkin was the first president of the International Union of Biochemistry and one of the most prolific, if not the most prolific, cataloguer in biochemistry. With Howard Mason as coeditor, Florkin prepared and published successively seven volumes of a treatise entitled *Comparative Biochemistry,* in the period 1955–64. Florkin not only made the point about unity in biochemistry but also explained in exquisite detail the need for energy in living things.

Proteins are large molecules, composed of smaller ones, amino acids. There are twenty kinds of amino acid in protein, linked in various ways. Typically, fifty to three hundred of these molecules are joined together in any one molecule of protein. The unity of biochemistry is again displayed by the fact that the relative quantities of these twenty types of amino acid are very similar in different organisms.

This recital of protein composition is linked to our questions about origins by the fact that all organisms display a *unity of biochemistry* in general and a unity of quantitative amino acid composition in particular. In the light of a wealth of solid empirical evidence for the unity of organisms at all levels of the evolutionary hierarchy, a similar composition for the first organisms seems to have been likely. At least this made reasonable the hypothesis that led to doing experiments with amino acids as a first step leading to organisms. It thus made sense to try to put amino acids of different types together to see what might ensue. Without the guiding principle of the unity of biochemistry in application to amino acids, the prospect for experimental retracement of molecular evolution would have been chaotic and negative. A faith in the orderliness of our universe at all levels, including that of amino acids, was also needed. The faith was subsequently supported by the experiments. Analysis of the lunar samples showed that the surface of the Moon contained no amino acids as precursors of protein but did contain solid precursors of amino acids. Our interest in what was to be found on the Moon was part of the question of where the amino acids exist as possible precursors to the protein.

Later chapters will explain the roles of DNA/RNA, phospholipids, and other biological substances, as well as a number of the many crucial functions of proteins. For the present, our question is, Where did these amino acids come from? What the amino acids, as building bricks for protein, were doing for original life will be taken up later.

The question where on the Earth are amino acids that could be converted to life is pointless today, since our planet is contaminated with amino acids *from* living things already. The Earth is contaminated in the soil (lithosphere), in the oceans, lakes, and rivers (hydrosphere), and in the air (atmosphere). All of these sources contain rich supplies of life-derived amino acids or of substances or organisms that can release amino acids. The question of prelife amino acids can be confused by the presence of life-derived amino acids.

For this reason, experts asked, especially in the 1960s, whether sets of amino acids are also present in extraterrestrial space. Since the 1960s, meteorites from outer space have been analyzed for amino acids. These have been found.[6] The first sets of amino acids, however, looked in the main like sets found in terrestrial organisms, for example, in the touch of a human fingertip. The general belief has been that these meteorites were contaminated by the fingertips of those who picked them up. Contamination of meteorites is likely also because they come to Earth through its atmosphere, which contains bacteria and their constituent amino acids. The soil is even more contaminated. The Russian microbiologist A. A. Imshenetsky, for example, placed a sterilized meteorite on the soil of Moscow and found it to be contaminated to its core within four days.

Because of the belief that all meteorites examined for amino acids were contaminated by Earthly sources, many investigators looked forward to analyzing lunar soil samples returned from the Moon by the Apollo astronauts. According to this reasoning, amino acids found in carefully handled samples from lunar sites would be free of the doubts that continue to apply to those found in meteorites.

Apollo Amino Acids as an Example of Channeled Thinking

The question of whether amino acids were indigenous to the Moon or the result of contamination deserved to be answered with utmost care and deliberation. Four teams were assigned to this question by NASA. Three of them did analyses for several kinds of organic matter, whereas our team concentrated solely on amino acids.

One team, which included Harold Urey, a doyen of planetary science, was headed by Bartholomew Nagy of the University of Arizona. Paul Hamilton, an expert on amino acid analysis, also participated. Another team was headed by Cyril Ponnamperuma; Charles Gehrke, who had long analyzed amino acids in foods and feeds for the Missouri Agricultural Experiment Station, was in charge of the analyses. John Oró of the University of Houston headed the third team, which did not report on amino acids. In fact, Oró finally functioned as an unofficial umpire on the amino acid question. The fourth team, my responsibility, included Kaoru Harada, an impeccable chemist, whose upbringing had the typical Japanese emphasis on cleanliness. That emphasis on hygiene was visible in his chemical work. The third member was Ed Hare, a chemist at the Carnegie Institution of Washington. Hare was, and still is, a leader in development of automatic amino acid analyzers. He had joined us years before the Apollo research, at his own request, in order to test his developments. While the other three teams were concerned with numerous problems in analyses for a variety of compounds from the Moon, our experience and interest were deliberately confined to amino acids.

Samples for amino acid analysis were expected from all of the Apollo landings. Six such landings occurred—they were Apollo 11, 12, 14, 15, 16, and 17. Collections from approximately ten sites were returned to Earth. The preemption of public attention occurred at the extensively publicized first meeting of the lunar analysts, in 1969 in Houston, following the first lunar landing.

The Nagy group and our team reported the presence of amino acids in the lunar samples.[7] The reports were tentative since both groups wanted to see how other, later samples would analyze. At this same meeting, one other group reported no amino acids in its analysis. This result, or nonresult, was widely reported in newspapers from Timbuktu to Tokyo. Once that occurred, much media and public perception was hardened in the belief that no amino acids were present.

However, Nagy's group and ours still wanted to answer the question thoroughly and deliberately. We examined samples from Apollo 12, 14, 15, 16, and 17, wherever on the Moon those samples came from. All of those samples, collected and returned carefully from over a quarter million miles away, yielded quite similar results. The analyses showed very small amounts of amino acids, but the analytical results were significant and repeatable. Also, contaminating sources like human fingerprints give nineteen or twenty amino acids, but these samples gave only the same five or six amino acids.

Neither we nor other observers were convinced by our first results, because the amounts found were minute, and we wanted to be especially sure that the amino acids were truly present in the lunar surface before the astronauts and their vehicles arrived at the Moon's surface—that is, that they were not contaminants. By using the special method our group had developed earlier, we obtained the same set of amino acids from lunar soil from Apollo 12 as from Apollo

11. This increased confidence in the analyses we had obtained for Apollo 11.

To scrutinize the situation even more closely, both Harold Urey and I asked for a "special environment sample" from Apollo 14. The sample was so designated because the astronauts were specially instructed on how to avoid contamination of the samples they collected. Several of the chemists who carried out analyses at the bench, including Harada and Hare, then met in one laboratory for analyses of those samples. Harada's cleanliness came to the fore; the analysts decided to let him prepare all samples of water. When scientists disagree on results, they tend to argue the issues in conference halls rather than to meet in laboratories for direct comparisons, so this entire event was somewhat unusual.

A closed conference was arranged in advance for a report on the findings from this sample. At that meeting, all hands reported essentially the analyses that had resulted from our and Nagy-Hamilton's examination of Apollo 11 samples.

In summarizing the overall Apollo results, a member of the group that did not find amino acids commented that their expectations for a "bonanza" of organic molecules had vanished. This was an allusion to expectations largely derived from experiments in closed flasks in the laboratory. Only a few shared the expectations, partly because there are no closed flasks on the Moon. At stake was more than just who had been right and who had been using the meaningful method of analysis. It also concerned such questions as how widespread are prelife compounds in the Solar System.

But the public perception, including that of most scientists, was already preempted by the well-publicized verdict of "no amino acids," as reported from the first Houston conference. From a "bonanza" of organic compounds, with emphasis on the most interesting units of protein, the pendulum had swung to no amino acids at all. Looking back, we can see that the choice of the method of analysis used was responsible for this later analytical judgment. The complex method employed in one of the laboratories had given many signals, most of which remain unidentified to this day. Among them, however, were the signals for five to six amino acids that our analysts had obtained from twelve separate lunar samples. Others, using a different, cumbersome method of analysis, were correct in saying at the invitational conference, "We have not yet identified with certainty a single molecule." For our method and results, Harada's clean sample preparation and Hare's sensitive sample analysis made possible a sure statement, one that other groups could echo. The other methods, however, did not permit making a leading statement.

The argument trailed off. Finally, in 1981, Oró with his coauthor, Seiji Yuasa, published an explanation of these accepted "well known results."[8] The yeomanly efforts of Harada and Hare at last received the reward of scientific acknowledgment, but it was belated and minimal. No reporters were present to hear and observe this later statement and to correct the highly publicized negative statement of eleven years earlier. The public was not brought up to date on a topic that was no longer news.

As Morgan had warned, the power of preempted perception is very influential.

The Preemptive Quality of "Origin of Life"

The phrase *origin of life* is itself a preempted perception. The subtle effect of this choice of words is to suggest that life came suddenly into being and was totally different from what preceded it. If that description of suddenness applied, life had an origin. Certainly, the concept of *origin* is reinforced by, and reinforces, the pentateuchal idea that life was created by a Divine Master who molded it out of clay.

If, however, as now seems scientifically sound, life is a result late in an evolutionary sequence that began with the Big Bang, then life emerged from some much earlier process that is more rightfully called the origin. With our present knowledge, it would be helpful to say *emergence of life,* a phrase that more accurately describes the modern, experimentally derived perception of the facts than does *origin of life.*

Alexander Oparin did more than anyone else to popularize the phrase *origin of life.* He was one of the first to perceive and write about the idea that life arose from a cosmic matrix. However, he did not see how such events occurred, and he personally embraced the order-out-of-chaos notion. Although for most of his life he was outwardly a supporter of Soviet ideology, which claims to be evolutionary rather than biblical, many Russians of Oparin's age had a deeply religious upbringing. It seems reasonable to suppose that Oparin's thinking about origins was channeled by his childhood society more than he knew.

Our understanding of what we know about life's beginning is similarly channeled. When the work of our laboratories began to interest newspaper reporters, I was frequently asked such questions as "When can we expect to hear that life has been synthesized in the laboratory?" This question still had biblical overtones. In an evolutionary context that, in true evolutionary style, depended upon stepwise processes, the reporter's question could be rephrased as follows: "At what step of the evolutionary sequence are you in your research on how life began?"

Even then, the question still had no single answer. It was possible to say early in the research, before 1960, that a kind of primitive, or primordial, cell had been produced by its precursor protein itself from the precursor protein. Since this step involved the production of organized microscopic units from an irregular powder in the ordinary microscope, it was a dramatic one. The process is known as self-organization. It was, however, possible to state that the resultant cells of this type, protocells, were distinguishable from modern cells primarily in that they contained no DNA or RNA.

It took the better part of twenty years of investigation to establish the capabilities and other properties of these laboratory protocells. The collective result is even more dramatic than the appearance of the protocell in the first instance. The proteinoid microsphere, as this model for the protocell is known in accepted scientific language, contains the discernible beginnings of all or nearly all of the salient properties of the modern cell—to some degree or in some precursor form.

This step from unorganized material to organized units is now recognized as crucial. Especially significant, in hindsight, is the appearance of an ordered set of cells and of the properties that these already possessed.

Much of the thinking on the subject of how life began has been channeled, for example, by mythological concepts. Even scientists, including those living in an officially atheistic state, such as Oparin's Soviet Union, appear to be influenced by the thinking of the past.

Scientific questions of life's beginning have been subject to such a channeling influence, much as Morgan observed for biology in general. Decisions had to be made on two chicken-egg questions. The answer to each had directive effects. One asked, Which came first, the cell or the protein? The other concerned molecular information rather than structure: Which came first, DNA or protein?

The Chicken-Egg Questions

PROTEIN FIRST OR CELL FIRST?

The first question was expressed best by Harold Blum of Princeton:

> I do not see, for example, how proteins could have leapt suddenly into being, yet both heterotrophic (external) and autotrophic (internal) metabolism are, in modern organisms, strictly dependent upon the existence of proteins in the form of catalysts (enzymes). The riddle seems to be: *"How, when no life existed, did substances come into being which today are absolutely essential to living systems yet which can only be formed by those systems?"*[9]

The solid structure of living cells, both in principal functions and amount, is made up much more of protein than of anything else. This is the reason for Blum's question. His comment is also due to a perception limited to protein and organisms.

An answer to this cyclic dilemma came by means of experiments that stepped outside the perception in the question. The experiments demonstrated that the first protein molecules could have arisen simply in the geological realm. The protein then assembled itself, as was also shown by experiments, into the first cells. At this point, the geological realm became the geological environment.

Many of the difficulties experienced by viewers of the proteinoid scene are rooted in their perception that the specificities—that is, the information that makes living things possible—arose within living things. When the perception is larger, for example, when the geological realm is seen as continuous with living organisms, the question can be rephrased and then answered. The impetus for rephrasing comes from the experiments that showed that proteins could arise geologically, that they did not have to be made by organisms. These experiments, using (geological) warmth instead of metabolism in organisms, were sufficiently successful for making proteins that the terminological authority, *Chemical Abstracts,* has indexed these polymers (large molecules of repeated units) as *thermal proteins* since 1972.

NUCLEIC ACIDS (DNA/RNA) FIRST OR PROTEINS FIRST?

A second chicken-egg question involves not protein and cells but protein and nucleic acid. Here, too, each is seen as requiring the other. Indeed, the specificities, or information in proteins, depend in each case on the exact arrangement of twenty kinds of amino acid in a molecule containing, typically, one hundred to three hundred amino acid molecules in a chain. The positioning of those amino acid molecules is dependent upon the arrangements of eight kinds of small nucleotide molecules in large molecules of polynucleotide, that is, DNA (deoxyribonucleic acid) and RNA (ribonucleic acid). So, nucleic acids are needed to make the proteins we know today. But to make nucleic acids (and proteins, too) enzymes are needed. No easy solution to the problem can follow inasmuch as almost all enzymes are proteins. Proteins are therefore needed to make nucleic acids, and nucleic acids are needed to make proteins. Which came first?

This problem is also one of perception. The two needs are not identical. Proteins are needed as catalysts to make large molecules; nucleic acids are needed to furnish blueprints for placing amino acids in their exact positions in specific protein molecules. Upon analysis of the problem, it becomes clear that it is not necessary to answer questions of equivalence. Instead, the first answer came from experiments that showed how specific arrangements of amino acids in protein molecules could occur in an utterly simple, geologically relevant fashion, without the complex modern mechanism.

Summary of Preempted Perceptions

We have reviewed preempted perceptions—that is, preconceptions—in the reporting on the question of amino acids on the Moon, in the effect of the phrase *origin of life,* and in the effect of the difficulties posed by chicken-egg questions of cell-or-protein-first and of nucleic acid-or-protein-first. Although all statements concerning this subject matter should in the best scientific tradition be viewed skeptically, we must be especially careful about negative ones, like those illustrated, because they tend to be self-fulfilling. If our preconception tells us that something can't be done, we don't try to do it. As the village caretaker said, "What hurts you the most is not what you don't know but what you do know that ain't so."

The two most effective preclusions in this area of research have been the result of preempted perception. The first is the dominant tendency of scientists to be reductionistic even when it is clearly shown that evolution itself moves in a constructionistic direction—that is, involves synthesis plus assembly. One cannot learn to bake a cake by "unbaking" one; the attempt to "unbake" destroys rather than instructs.

The second preclusion is the related tendency in science to extrapolate backward from sure knowledge to the original system. In this sort of extrapolation the practitioner learns a great deal, for example, by comparing fossilized organisms in ancient and modern geological strata. However, it does not permit traversing the boundary between the first organized cell and the prior unorganized precursor matter. The march of evolution depended upon nature's experiments in a forward direction. Movement in that direction has its own set of principles. It was necessary to retrace that movement. To learn how living units began, we needed to carry out such retracement in a context of going "Back to the Future"! The value of a forward-looking perspective resided additionally in a recognition that with evolution, as with a tree, how it began had much to say about how it developed.

By 1963, the chicken-egg question of "DNA/RNA first or protein first?" had been recognized. A few leading researchers looked at both possibilities, having mostly not made up their minds. The essential reason that one program continued in the protein-first paradigm is that it led to results that in turn generated new experiments and also led to new results. This is the pragmatic criterion for meaningful research, as Steven Weinberg stated in the passage quoted in the preface. However, with the focusing of thinking that typically comes from close attention to any one point of view, the intrinsic defects of the DNA/RNA paradigm began to be apparent.

The DNA/RNA paradigm has not led to new results, almost to no results at all. Questions about how DNA or RNA arose without prior proteins or about how these agents participated in the origin of biological information have received neither an experimental demonstration nor a cogent theoretical explanation. The DNA/RNA emphasis also fails to yield an understanding of a smooth transition of such compounds to cells, which are the units of life for all wide-angle biologists.

Moreover, the DNA/RNA mechanism as we know it is extremely complex. That such a complex mechanism could have been present at the beginning of organic evolution is to even the most imaginative scientist essentially unfathomable or very improbable. The alternative is that the complex templating mechanism (see chapter 7) that uses nucleic acids was preceded by simpler mechanisms. That answer brings the proteins into purview as indeed the logical simpler mechanism. The counterargument is that the nucleic acids that are now largely or entirely the repositories of biological information originally played that role. This argument typically fails to recognize that both nucleic acids and proteins are informational, although they convey different kinds of information. Accordingly, either primitive nucleic acid or primitive protein might have served as the initial source of information.* DNA contains hereditary information, whereas proteins contain kinetic information. What the research has shown is that initial proteins obtained their information not from the outrageously complex modern templating mechanism but from the simple reactions of amino acids to form the first proteins. The amino acids in sets contained information; the resultant thermal proteins contained information derived therefrom. Moreover, separate experiments have partially shown how the information in thermal proteins could later have been transferred to the first DNA/RNA.

When life began, there could have existed no logical analysis of alternatives and there could have existed no scientific plan for investigation. The beginnings were totally experimental in a natural sense. The information was in the molecules. A thesis of this book is that employment of the constructionistic approach in the laboratory is crucial, as it had been in the geological realm. Construction is experimental. This time it is man-made.

The uniqueness of the constructionistic approach has been brought out by Richard S. Young, who was a primary initiator in NASA's involvement in laboratory studies of the emergence ("origin") of life. He has emphasized the need for experiments that carry amino acids to proteins to cells—that investigate "the transition of the inanimate to the animate. I would argue that it is only in this fashion that the total story can ever be unraveled. No amount of

*The information theory specialist defines the term *information* in complex mathematical ways. We use the word here to mean any system that has the capacity for selective (preferential) interaction with another system. A molecule is a system.

dissection of contemporary cells will ever unveil this phase of evolution, nor is it likely to be found in the fossil record."[10]

We shall later consider the problem of life's beginnings from the perspectives of original locale, beginning compounds, more fully the chicken-egg problem of nucleic acids or proteins, and the beginning biofunctions, investigative approaches, and partial paradigms that are clusters of theorems. Our present focus is on what has proved to be a watershed question—the investigative avenue.

A phrase popular with young people in the generation of the eighties is "go with the flow." Whatever its origin, the phrase applies with full force to experimental attempts to retrace the pathway of evolution, instead of investigating analytically in the opposing direction. "Back to the Future" and "go with the flow" thus both express the young, new view.

So, back to the future and go with the flow. In more traditional scientific phraseology and syntax, the only known procedure for validating the course of early evolution is to retrace the steps in the direction of primordial to modern. This time, however, even though no teleological goal is involved, we know where evolution was going. As the analyst feels secure in confining his attention to what is here, so the evolutionary synthesist feels secure in the knowledge that retracement of past evolution must give him products, processes, and principles that need sooner or later to be discerned in that which is here. Knowledge of what is here is an accumulation for which the synthesist must thank the analysts.

CHAPTER 2

Beyond the Power of Science?

When a substantial proportion of society looked upon research on the origin of life as either sacrilegious or scientifically hopeless, many scientists within society inevitably shared such feelings, or were similarly affected. The preconception that life's emergence was miraculous was maximal about 1950, at the time that research began in our laboratory.

In those days, I was privileged to credit Harold Urey in a scientific paper for having had the courage to suggest related experiments and to let his student Stanley Miller perform them. Professor Urey, who was widely honored for the discovery of "heavy water," was indeed a courageous scientist. The Urey-Miller experiment yielded amino acids under conditions then believed to be early geological. This belief has not stood the test of time.[1] The change in belief gradually occurred as the opinion became more widespread that the reactant atmospheres employed had not been geologically valid and because it strained the imagination to suppose that laboratory apparatus could simulate structures in the geological realm.

Also, the making of amino acids is not equivalent to the making of life, although that concept was promoted by association, a kind of extrapolation favored by some popular science writers. Indeed, sets of amino acids were subsequently found in samples from meteorites and returned lunar samples alike, but the presence of indigenous life was rigorously ruled out for both sources. But to claim that such nonmiraculous chemical reactions could be studied as a scientific staging area for this purpose was a bold step on Urey's part. It made exploratory experiments in this area much more open to the perception that they could be studied. One could object that Herrera had earlier prepared the thinking more completely. However, the fact that a well-known scientist such as Harold Urey embarked on this study had far more impact than the pioneering work of a little-known investigator in Mexico City, described later on in this chapter.

Despite Urey's influence, many scientists to this day tend to avoid a question that to them seems beyond answering. In a purely scientific context, they correctly see life as extremely complex, whether or not they think of life as the result of a miracle.

In science, the *forward* approach to evolutionary problems is still unconventional to the point of being novel. The dissection of a chemical compound or an organism to study its components is regarded as "acceptable." Putting together a protein molecule or an organism has for many scientists appeared so unorthodox as to be sacrilegious. A few examples from the not too distant past illustrate this point.

Are Proteins Sacred?

About 1963, I found myself talking to a local section of the American Chemical Society in the mountain West. At a nearby university was a protein chemist, known to me for many years from his reports on analysis of protein. I was glad to meet him in person. By the time my lecture and discussion were concluded, it was clear that my colleague was not eager to meet me. The audience discussion after the lecture focused more and more on the numerous complexities of protein molecules. Finally, the local colleague could contain himself no longer. He rose in his seat in the auditorium and in a robust voice said, "Only God can make a protein!" Rational discussion was cut off. One of the friends of the protein chemist led him away by the arm.

Since that encounter, Bruce Merrifield at Rockefeller University and others have made a number of proteins. The Merrifield method however makes proteins in a way that obviously could not occur spontaneously in a geological setting. It essentially yields proteins such as have already evolved; it does not challenge the possibility of an original spontaneous generation of proteins.

But in the early 1960s I learned how emotionally bothered some chemists can be by attempts to produce complex, life-approximating molecules under geological conditions, while at the same time they feel comfortable analyzing the molecules that can be ascribed to Divine Providence.

A number of similar incidents, not all of them concerning scientists, come readily to mind. One of those involving nonscientists happened at about the same time as the Colorado lecture. The participants were two authors, Philip Wylie and Robert Frost. Wylie wrote popular articles and books. Perhaps his best-known book was *Generation of Vipers.* Robert Frost is such a popular poet that he needs no introduction.

Both Frost and Wylie lived in the same section of Miami as my wife and I. Wylie had a few years earlier visited our laboratory in Tallahassee, and he had recently stopped in again to look through the microscope at our new location in Coral Gables. He subsequently described what happened after he left our lab and met Frost near his home.

Wylie, in a not uncharacteristic burst of enthusiasm, related to Frost what he had just witnessed under the microscope an hour earlier and explained that these experimental results seemed to be a scientific equivalent of Genesis. Wylie reported that Robert Frost "fought it every inch of the way." Frost's first response was to regard such research as sacrilegious.

Wylie had evidence that the discussion was much on Frost's mind for the next few days. Two evenings later, Wylie and Frost met again, this time at a neighborhood gathering. Wylie entered the hostess' front door and found himself looking into the face of Frost, who sat in a big easy chair. On each arm of the chair was draped a "sweet young thing," while a third sat on the floor. All three were looking at the poet adoringly and engaging him in animated conversation.

As soon as Frost spied Wylie entering the door, he abruptly broke off the conversation with the young ladies and, just as abruptly, picked up the front-yard conversation of two days earlier as if he and Wylie were merely continuing it. Frost pointedly intoned, "Nothing is forbidden in research."

In addition to the effects of long conditioning by the pentateuchal explanation of the origin of life and to the fact that only analysis of God's handiwork is fully "respectable" to some scientists, other perceptions enter at the research level. These mingle with and reinforce those already described, especially for the most cautious scientists.

Such attitudes derive from the reality that scientists are conditioned by the rules of the game that require them to defend every advance they make. For example, the final oral examination of doctoral candidates is known as "defense of thesis" and is said to be "critical." A thesis that passes this kind of test is "sound." The most successful of the candidates are those who anticipate criticism and prepare to answer it in advance, either by careful argument or by additional experiments. This heavy emphasis undoubtedly makes for a sounder body of scientific knowledge, but at the same time it produces scientists noticeably more defensive in personality than practitioners of other occupations.

The problems of perception apply to origins as well as to evolution and to both in concert. The most celebrated case to arise from this situation was the Scopes trial, dramatized in books, plays, and motion pictures.[2] It concerned Darwin's theory of evolution, since only evolution was in the eyes of the educators and the public at the time. A scientific explanation of the initial creation of life had not yet appeared.

In the United States in 1925 the most widely accepted explanation for the diversity of life on this planet was the biblical one. The competing view, that of Darwin's origin of species, had been introduced in England approximately half a century earlier and was taught in biology classes in this country in the early part of our century.

This situation raised various questions. For example, should a high school teacher in Tennessee, John T. Scopes, be punished for teaching the theory of evolution, since antibiblical teaching violated state law? This question received international attention in a historic courtroom trial in Dayton, Tennessee, in 1925. Clarence Darrow, a "liberal," scholarly attorney, defended Scopes, while a silver-tongued orator, William Jennings Bryan, was attorney for the prosecution. Scopes was convicted and fined $100, but this verdict was later replaced by an acquittal on appeal. Bryan was not given time to deliver a speech on the sanctity of the Bible; at the conclusion of the trial, he became ill and died. All this drama occurred within the single, hot month of July 1925.

Forty years after the primary event, I found myself fifteen miles away in Knoxville, Tennessee, to deliver a lecture on the prelude to life—its protobiological, chemical beginnings. This presentation was one of a number of tours through clusters of states for diverse sections of the American Chemical Society. Between 1965 and 1975, I responded to requests from approximately ninety sections of the society to describe chemical experiments in the polymerization of amino acids and the selforganization of the products to form protocells. As always, the announcement of the evening's subject for Knoxville drew a large audience, even though the topic was far from as fully developed then as it was to be by 1980.

Beyond the Power of Science?

After I concluded my talk, the chairman of the Department of Chemistry at the University of Tennessee, who was also chairman of the meeting, surveyed the well-filled auditorium and said, "I have been looking to see how many of the local fundamentalists are here. I don't see any. I have to conclude that they aren't sufficiently well informed to recognize that now they have to fight the battle at the molecular as well as the biological level."

Those who interpret the Bible literally think that the question of the origin of life belongs only in the realm of the miraculous. For my part, when I set out to understand how protein might have arisen suddenly, I did not imagine that the consequences of that search would take me partly out of the scientific realm. Both outside and inside of science, the problem appeared to be "insuperable," to use the scientist George Wald's word in his 1954 assessment of it. As we shall see, the reasons for its insuperability in these two realms were different, even though the reasons might overlap in individuals. For a scientist, however, no defeat is as complete as the conviction that it is hopeless to try.

Of the persons who have interpreted the Bible figuratively, one of the most up-to-date and thorough came out of a small denominational college in Texas in the late 1970s, when I led a number of two-day seminars on the origin of life for the American Association for the Advancement of Science. Each class was limited to twenty-five advanced teachers. The students of one of the teachers reporting in the class in Austin, Texas, had taken the initiative in producing a movie on Creation Week. They organized it approximately as follows:

Monday	DNA
Tuesday	RNA
Wednesday	Protein
Thursday	Lipid
Friday	Cell
Saturday	Metabolism
Sunday	Rest

Any attempt to imitate in the laboratory the making of a first cell had to reckon not only with literal and figurative interpretations of the Bible but also with a formidable scientific stone wall. The cell, the unit of life as we know it, is by scientific analysis terribly complex. This complexity was, in the early part of this century, enough to scare off almost all optimistic experimenters. In the light of proper evolutionary emphasis on stepwise processes, one might more logically have chosen to produce instead a primitive cell, because it was simpler than a modern cell. Ironically, one would eventually have learned that primordial cells, protocells, are in some ways more complex than modern cells. The defeatist attitude, however, was conditioned by the recognition of the tremendous complexity of each modern microscopic living unit. But that complexity should not have been the problem that delayed research. The more important consideration was the complexity of the *processes* that yielded a first cell. We now have many reasons to believe that the product was maximally complex while the processes were maximally simple in operation. The complexity of the modern evolved product makes it easy to understand the assumption that the initial process had likewise to be highly complex, but research has shown that this transfer of judgment from product to process was not justified.

The process proposed as having produced the first cell is in fact simple, comparable to the mythological simplicity of Adam's having been molded out of clay. In light of the experimental discipline now available, it seems obvious that a first cell on Earth, as lifeless and devoid of human intelligence as Earth had to be, would have had to arise by simple processes. That idea was not prevalent in science, however, largely because almost no one thought about the problem in such evolutionary terms.

If anyone had cared to reason in an evolutionary context, the problem might have been solved much earlier. We now know that simple arrays of components can interact to lead rapidly to complex products. In mathematical terms, this transition from the simplicity of a few elements at the beginning to complexity is known as exponentialization. For example, the use of only four units—A, B, C, and D—can produce 24 different arrangements.* If the number of units is increased to ten, the number of potential arrangements jumps to 3,628,800. When one combines the concept of exponentialization with the evolutionary power of stepwise reactions, simplicity can and does easily meld into complexity that permits a huge variety. A small number of organic components can thus give a great variety of living assemblies.

I hasten to add that this reasoning occurred *after* the creative act in the laboratory, not before it. The objective in our laboratory was to make a *protein* as the geological realm might have done it. Only after the experiments had yielded that protein material did we have reason to think about how it might have turned into cells. The enabling thoughts thus themselves evolved.

The new vista that resulted in this stepwise thinking came at an important turn in the research road—during my attendance at the romantically named Seaweed Symposium, in Galway, Ireland, in 1958. We had been brooding about how proteinoid would have been converted into a primitive cell. At the meeting, I found time to activate research on the problem. Such planning in scientific research, incidentally, is often done at sites away from home. Far from the harassment of committee meetings and the frequent jangling of the telephone, I found I could think clearly about the problem. Allowed the luxury of contemplation, I was able to send the following note from Galway to Jean Kendrick, a skilled technician in my lab in Tallahassee:

*ABCD BACD CABD DABC
ABDC BADC CADB DACB
ACBD BCAD CBAD DBAC
ACDB BCDA CBDA DBCA
ADBC BDAC CDAB DCAB
ADCB BDCA CDBA DCBA

14 Aug 58

Dear Jean:

There is one more experiment I have been ~~thinking~~ of in preparation for Oct., and which I would like you to place on high priority, as follows:

Make proteinoid (2:2:1) in usual way (2g. or so enough) lift hot melt from oil bath with clamp and immediately add to tube 20 ml. of boiling H_2O (Protect eyes!). When this settles down, swirl, pour off supernatant while warm. When cool, examine this under microscope. ~~If it~~

If there is anything cell-like or globular repeat with proteinoid-nucleicoid when and if latter is obtained and characterized.

Sincerely
S. W. Fox

Figure 2.1

Beyond the Power of Science?

This incident illustrates how much research gets done by skilled assistants who devotedly act on the various ideas and plans provided by someone who has research funds to pay them. It is usually a happy combination. Mrs. Kendrick was productive and knew it.

Before it was clear that the origin and evolution of life involved a stepwise approach, other researchers had dealt with the problem from their own perspectives. For example, Alfonso L. Herrera (1868–1942), mentioned earlier, was a scientist of great breadth and optimism. The originality of his research and thinking may have been aided by the location of his Laboratory of Plasmogeny in Mexico City. To be off the beaten path scientifically, it often helps to be off the beaten path geographically. Mexico City has, of course, long been one of the great capitals in the world, but in the late nineteenth and early twentieth centuries it was rather far from the scientific mainstream. Herrera thus had the benefit of an "ivory tower" location in which to do pioneering work; on the other hand, his work received inadequate attention.

By the 1930s, Herrera had easily and quickly produced organized microstructures that had the look of a great variety of cells. Moreover, he started many of his experiments with substances such as formaldehyde and ammonium thiocyanate. Decades after these experiments, new investigators found that the materials he used were organic substances abundant in interstellar matter, that formed in tremendous quantities around the stars. Herrera chose substances prominent in the limited number of types in interstellar matter. In this, he was far ahead of his time.

Herrera's contributions were attacked quite unfairly. Alexander I. Oparin, the Soviet biochemist who developed an extensive following through at least seven books on the origin of life, belittled Herrera's results. Oparin claimed that the resemblance of Herrera's units to living organisms was no greater than the "resemblance between a living person and a marble statue of him."[3] He also asserted that Herrera referred to his own results as "a means of obtaining living organisms artificially."

This attack by Oparin undoubtedly contributed to negative reactions of others, even in Herrera's own country of Mexico. But Oparin's statements are not in tune with the facts. Herrera's last article, in 1942, later cited by Oparin, emphasized advances in theory and stated, "The particular theory offered here lacks confirmation."[4] Hardly the dogmatic tone imputed to him by Oparin! Herrera's experiments were in a forward evolutionary direction, like the Urey-Miller experiment, but without the benefit of the more elegant analyses Miller was able to do in 1950. Oparin's experiments in the production of artificial cells, however, were actually far behind Herrera's. Oparin and his colleagues used polymers from already evolved organisms to make a kind of cell known as the coacervate droplet, whereas Herrera used materials that might indeed have served as precursors in the progressive evolution to cells. Oparin's gum arabic and gelatin could not do so. A face-to-face conversation with Oparin in 1979 through an interpreter satisfied me that by that date Oparin had come to recognize the inadequacy of his own laboratory model. Oparin stressed to me, however, that coacervate droplets could reveal principles of cellular organization, even if they did not specify how cells first arose. I agreed with him in this last comment, a few months before he died, and emphasized that they served this role very well and early in his research.

A. I. Oparin and His Coacervate Droplets

Since the cell is the unit of life, the protocell is the unit of protolife. In the textbooks, the two most mentioned protocells to have been modeled in the laboratory are Oparin's coacervate droplets and our proteinoid microspheres. In fact, Oparin once told me, in response to a question about this way of looking at the two types, that he considered proteinoid microspheres to be a kind of coacervate droplet.

Years later, I disagreed with his answer (I am sometimes slow at making up my mind). The differences between coacervate droplets and proteinoid microspheres have been pointed out in a number of up-to-date textbooks. The main differences are in stability, in regularity of size, and in ability to serve as a replay model of the emergence of the first cell on Earth.

The criterion of stability is best illustrated by another personal experience with Oparin. As we noted earlier, Oparin popularized the chemical perspective on the origin of life on Earth. He ranked very high in the councils of those who advised the Soviet Presidium on science. For many years, he was director of the Bakh Institute of Biochemistry in Moscow, which specialized in practical aspects of fermentation, of wine and beer especially, and in the theory of the origin of life.

Except for the making of coacervate droplets, Oparin's contributions to an understanding of the origins of life were dialectical, that is, "theoretical." Carrying on discussions requires nowhere near the time, physical effort, or money for apparatus and technical help that an experimental program demands. Everyone agrees that the experimental frame of mind and experimental results can make unique contributions. They sometimes teach us to think in ways we would not otherwise have discovered.

Albert L. Lehninger, a famous faculty member at Johns Hopkins University, in his textbook *Biochemistry,* emphasized in 1970 the importance of experimentation:

> Not too long ago inquiry into the origin of life was considered to be a matter of pure armchair speculation, with little hope of yielding conclusive information. But many scientific advances made in the last two decades support the view that valid answers to some of these fundamental questions can be deduced. . . . In this final chapter we shall survey some of the experimentation. . . .[5]

Lehninger also stated, "Oparin's coacervate droplets are obviously models; they were made from biologically formed substances and were not formed under simulated primitive-Earth conditions . . . they do not provide for evolution."[6] Despite Lehninger's negative remarks about coacervate droplets, Oparin was delighted to see Lehninger's chapter in print. In general, I think Oparin derived an unqualified joy from experimental advances.* He could always feel a grandfatherly relationship to any new findings.

How Stable Are the Protocell Models?

The aspect of stability of the two protocell models was brought into sharp relief when I visited Oparin's laboratory in the Bakh Institute of Biochemistry in 1969. The national academies of the two countries had arranged a lecture exchange in which the Oparins were to spend a month in the United States and I was to travel with my wife through the Soviet Union for a month to speak in Moscow (twice), Leningrad, Kiev, Tbilisi, Yerevan, and Tashkent. With the exception of Yerevan, where a brilliant Armenian, H. C. Bunitian, was in charge, the quality of science declined as one went farther from Moscow. So did the quality of translation. Since my wife speaks Russian, we could judge that from the questions asked.

The translation in Leningrad was particularly disappointing. When I told the audience that people in my laboratory had expended 120 man-years in research on the problem, a Russian in the audience exclaimed, "That can't be; he doesn't look anywhere near that old!" Of course, I had been unwise to use an idiomatic expression to suggest how much work had been done. I also had the pleasure of being introduced in the Soviet Union as a speaker from the Institute for Molecular Revolution. In reality, molecular evolution and molecular revolution are not so far apart!

On this lecture exchange, Oparin arranged for us to stay three days

*Some theorists prefer the belief that the emergence of a cell in any form was deferred. This has a seductive, Aristotelian logic to it in that *modern* cells are complex beyond their components. The *experiments* that mimic natural evolutionary experiments indicate that a cellular membrane could have arisen quickly; the derivative reasoning suggests stepwise cellular evolution. An incomplete cell (a protocell), according to this, arose and evolved to a complex modern cell through a long series of changes. Indeed the study of cells that we can examine now lines them up in stages from simpler structures to more complex.

in Moscow before the next visit, to Leningrad. Earlier, in 1963, when Oparin had attended our international conference at Wakulla Springs, Florida, I had made proteinoid microspheres for him in our university laboratory and given him samples of proteinoid that he could take back to Moscow to make his own microspheres.

Now, in Moscow in 1969, he was returning the favor. We went into the laboratory, where his assistant produced some coacervate droplets. These are usually made from gum arabic and gelatin. Gelatin is a product of the partial breakdown of collagen, the most abundant of proteins in mammals, including humans. When they are brought together in water, the oppositely charged colloids neutralize each other and form a new compound, known as a coacervate. The coacervate separates into droplets. When the technician in Oparin's laboratory dipped her glass rod into the test tube and then used the rod to dab coacervate droplets on a microscope slide, there was nothing on the slide except a track of a few oily droplets of no particular shape. Her and Oparin's disappointment and embarrassment were obvious. Two other attempts failed similarly. I was also quite embarrassed for them, even though I was already aware that the process was not invariably dependable. This unreliability, however, did not harm Oparin's thesis seriously.

Oparin then telephoned a favorite former student of his, Tatiana Evreinova, at Moscow State University (MSU). I had already been scheduled to stop at MSU that afternoon. When I visited Evreinova there, she made some coacervate droplets for me, with a little more success, but also with some evident disappointment. She then brought out some coacervate droplets on a slide, which were already brilliantly stained and had been fixed (see fig. 2.2).

The technical point that this visit graphically drove home to me is that the coacervate droplets have little stability. Oparin was frank about this in his writing. Using the coacervate droplets to guide his thinking, Oparin admitted the need for increased stability. In order to explain how this stability might have been attained by an evolving prelife, he invoked Darwin's principle of natural selection. His 1957 book dealt at length with this problem.[7] Those coacervate droplets that happened to have greater stability than others and could be

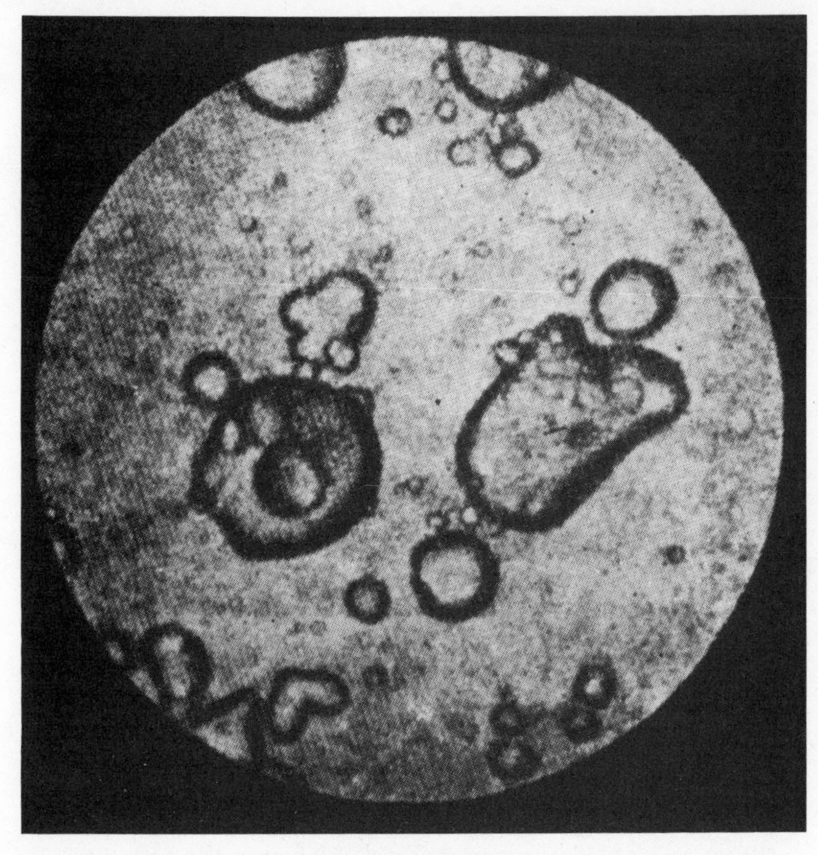

Figure 2.2
Coacervate Droplets

Coacervate droplets by Evreinova. These are composed of gum arabic, gelatin, and RNA. The coacervate droplets lacked stability, as Oparin recognized. Since they are composed of modern components they cannot represent evolvable protocells, as Lehninger emphasized. They fail to reveal the generative significance of the self-ordering of amino acids.

reproduced would, according to Darwinian principles, come to dominate the weaker ones. Thus, the earliest cells would become more stable by natural selection.

For at least twenty years, the idea of a need for evolution by natural selection to develop cellular stability influenced thinking in the field. George Wald expressed it in his classic 1954 paper in *Scientific American* on the origin of life. Wald cited the gradual development of the first cells through stages of precellular aggregates of protein.[8] When NASA officials asked me to organize an international conference for 1963, I entitled it "The Origin of Prebiological Systems" for the same reason.[9] We all visualized biosystems as prebiological because Oparin had analyzed the problem that way. Channeled thinking!

Oparin felt that proteinoid microspheres (see fig. 2.3), unlike coacervate droplets, were too stable, because he equated that stability with a lack of dynamism. The proteinoid microspheres have, however, since been shown to be extremely dynamic—far more biochemically dynamic, for example, than any other protocell model yet produced. Their stability had by 1975 been shown to be such that they could last indefinitely. The microbiologist Laura Hsu made some under sterile conditions. Normally, microspheres exposed in suspension serve as food for bacteria and fungi from the air. Since Hsu's batch was protected from such contamination, the microspheres lasted well—in fact, for six years. They were then discarded, their durability and stability having been established. Natural selection was not needed to improve stability. On the primitive Earth, proteinoid microspheres could have waited indefinitely for the next events.

Oparin's reliance on natural selection to overcome the stability problem led him to regard his microscopic units as *pre*cells and to postpone the problem of protocells. In other words, he assumed the existence of cell-like precursors to the first cells and avoided the thought of stable protocells. The precells could, he reasoned, develop stability through the descendant generations, which were possible theoretically because the droplets could reproduce. Even oil droplets will divide in two. According to Darwin's selection principle, daughter droplets that happened to have greater stability than others during the course of natural variation would be more likely to survive. Increased stability would be inherited, although how inheritance would work for coacervate droplets was a question that was not addressed. The quality of stability would be bequeathed to the next generation,

Figure 2.3
Proteinoid Microspheres

Scanning electron micrograph of proteinoid microspheres. Note uniformity and huge number. These are nearly always of a diameter size of 1–3 microns. Illustration prepared by Mr. Steven Brooke of Steven Brooke Studios.

which could also yield increasingly stable offspring droplets in the same way. Finally, according to Oparin's argument, cells stable enough to start an evolution would arise. Oparin's special view was to dominate for two to three decades theorists' understanding of the origin of life. To recapitulate, precells evolved to protocells by natural selection, serving to preserve increasingly stable precells.

Invoking natural selection at the precellular stage was indirectly necessary because, as it turned out, the wrong experiments had been picked to represent what could have happened. Experiments can be uniquely educational, but they are best when correctly chosen.

Although Oparin was mainly dialectical, he did a few kinds of experiment. Nearly all of them were on coacervate droplets, which he perceived as representations of protocells, but he also repeated with others our report on the polymerization of mononucleotides to polynucleotides by the use of catalytic proteinoids.

Oparin was among the first to understand and talk about the importance of the astronomical and geological prelude to living things. Ironically, however, he failed to apply this awareness to the cellular simulacra with which he worked. We have here a striking example of the remarkable degree to which the use of an experimental model influences the theoretical musings of the experimenter.

Oparin employed in his laboratory studies materials that are found in the cells of modern organisms—gum arabic (from the acacia tree) and gelatin (from the hoof of a horse).[10] These are substances from a highly evolved plant and a highly evolved animal, respectively. Had he followed in the laboratory his own theoretical perceptions, he would have used materials that had arisen out of geological precursors preceding cells, but he did not have a model of those.

Biochemical logic and the experiments tell us that chemical precursors to cells are proteins. The biochemical precursors to proteins are amino acids. The logical chemical links in an evolutionary sequence between the geological rain and the prebiotic proteins would thus be sets of amino acids. But Oparin used evolved, already formed, proteins from evolved organisms.

That choosing the wrong material deludes the experimenter is further illustrated by the instability of these cells. The stability of coacervate droplets made of gelatin and gum arabic is of a low order. In some cases, the products in the laboratory are more like loose oily droplets than like cells.

Some observers find that Oparin's reliance on natural selection as a primary force in evolution was consistent with Soviet ideology. Natural selection gives to the environment (the state) the primary power. The idea of molecular selection as a primary force, while not denying environmental influence, honors more fully the intrinsic, or internal, worth of the individual. Indeed, the first "individual" emerged from its surroundings, which could only then be called an environment. While it is true that the individual first emerged as part of a social group (see chapter 4), the latter is composed of individuals. Individuals and the group are in a sense not separable. However, whether the initial force is from the outside, as in natural selection, or from the inside, as in molecular selection (see chapter 8), can be judged from the model. The self-ordering of amino acids (molecular selection) can be even more effective than genetics in its emphasis on control from the inside.

Attractive and perhaps perilous as it is to extrapolate from the behavior of primordial cells to that of political humans, can one even do research on the origin of cells? Earlier we alluded to the thinking of the poet Robert Frost, who came to believe, after soul-searching, that no choice of research should be denied. But there are Ph.D.s in science, and other people as well, who believe that research on how life began should not be permitted. These Ph.D.s are not the same as those scientists who believe, for various reasons, that such studies are too scientifically hopeless to be encouraged.

The most religious fundamentalist scientists have become known as "scientific creationists." As they have met substantial opposition from the laity and the clergy alike, they have tended to change both their names and their tactics. One formal group, the Foundation for Thought and Ethics, in Dallas, has had a vogue. It has been headed by a doctoral graduate in chemistry from Iowa State University,

Charles B. Thaxton. In the book *The Mystery of Life's Origin,* Thaxton, Walter L. Bradley, and Roger L. Olsen collected many published papers that they considered attempts at a scientific understanding of the origin of life.[11] They then criticized each of the reports, frequently relying on negative comments by other "origin-of-life" researchers. In so doing, Thaxton and his coauthors generally did not report published responses to these criticisms. These one-sided treatments are reflected in the summary in their book. In that summary, Thaxton and his coauthors presented five alternative judgments for the reader:

1. New natural laws are needed.
2. Panspermia (life came from elsewhere).
3. Directed panspermia (a suggestion by Crick and Orgel that life was brought to this planet purposely by spaceship).
4. Special creation by a Creator within the cosmos.
5. Special creation by a Creator beyond the cosmos.

Alternatives 4 and 5 are obviously creationistic—5 is more so than 4, since 4 comes closer to the scientific view that the cosmos is a product of its own earlier activity.

The authors concluded that an answer to the question of whether an intelligent Creator did create life is beyond the power of science. However, questions about the Creator are theological, not scientific. Thaxton's alternatives 2 and 3 are really a kind of science fiction.

Only alternative 1 can be referred to as scientific. There, as in the others, Thaxton is requiring that the scientific answer meet the standards set by Scripture. The listing of the item itself implies that the present laws of science are inadequate. This view is based on the assumption, widely held both by true scientists and by creationists, that life arose as the result of chance events (random matrix and random occurrences). If new laws were required, as Thaxton and others claim, all the alternatives, as they define them, are now outside the realm of science. But we shall see that the established scientific principles are good enough when they are disciplined in the evolutionary directions of inside to outside and of back to the future.

Many of the tactics of the earlier "scientific creationists" in the San Diego group, including Henry M. Morris and Duane Gish, were similar to those of Thaxton and his coauthors. They set the Bible as a standard and did not recognize that the competing evolutionary explanation has its own guidelines. In their book *Scientific Creationism,* they asserted that creation "is inaccessible to the scientific method" and that "it is impossible to devise a scientific experiment to describe the creation process."[12] This claim is correct if one limits the scientific possibilities to instant creation. In evolution, stepwiseness is a required enabling theme that avoids such paralyzing limitation.

The scientific credentials of the leadership of the Institute for Creation Research in San Diego are quite weighty. The director is Henry M. Morris, who earned his Ph.D. in hydrology from the University of Michigan. Morris has written a religioscientific treatise on The Flood. One would probably search in vain for someone with a broader background of information to write on the biblical Flood!

Duane Gish has very strong scientific credentials. As a biochemist, he has synthesized peptides, compounds between amino acids and proteins in size. He holds a Ph.D. from the University of California at Berkeley. He has been coauthor of a number of outstanding publications in peptide chemistry. This background coupled with critical perceptions about evolution, which appear to be more accurate than those of many neo-Darwinists, and excellent debating skills have made him a leader in the public contest between evolutionists and creationists.

Morris and Gish do deserve attention in a scientific framework for their arguments that evolution based on a random context is indefensible. This one criticism by them is sound, even though they wish to overcome it by introducing the determinism evident in modern living forms; they do this by invoking divine action.

The "scientific creationists" emphasize the incompleteness of the fossil record, but they do not comment on the microfossils of Elso

Barghoorn (1962) and those of J. William Schopf or on the proteinoid microspheres (1959 and later), which the microfossils resemble. They also fail to mention Herrera. Their continued emphasis on the complexity of living systems suggests that the San Diego group feels that cells are so complex as to be "out of sight," a view capable of eliciting sympathetic accord from even many nonreligious scientists. The experimentally demonstrated simple emergence of a protocell, followed in evolution theoretically by multiple steps to a modern cell, is an alternative they do not consider in their text.

A focus on this group's criticism of prebiotic proteins shows in tandem three especially false statements: that laboratory conditions would not have existed on primeval Earth; that the proteins produced were not ordered, biologically useful, or chemically specific; and that the proteins would have quickly been destroyed.

These claims, also made by noncreationist scientists, are readily answered. The conditions needed to make these substances are widespread on Earth, even today. Many papers from numerous laboratories show that the thermal proteins are highly ordered, although this fact continues to surprise other scientists. Finally, the book *Scientific Creationism* cites as support for the third point a paper from our laboratory, in which the evidence is quite the opposite of that stated!

Michael Ruse, an evolutionary philosopher, in his chapter in *Science and Creationism,* refers to "the Creationists' lies and underhand tricks." He characterizes debates of true scientists with the "scientific creationist" as "circuses."[13]

But we can ignore much of this brouhaha and look for a "bottom line"—namely, the need to explain the limited but extensive variety of living things and how they came to be, right from the first organized cell. That living things are limited in a special way is seen in the similar appearances of members within modern human families, in the fact that all dogs, cats, and raccoons are quadrupeds, have tails, and so forth. That the families exist in an abundant variety is also self-evident. These similarities of type are all manifestations of a substantial degree of determinism.

This biota either came to be by chance, that is, randomly, or it was directed. If it was directed by a clay-molding Intelligence, that explanation, with auxiliary concepts, would meet most of the requirements of a highly deterministic, modern world. Some scientists, notably many neo-Darwinists, believe that the necessary determinism was introduced along with DNA and that, in fact, life did not begin until DNA emerged. Others hold that all evolutionary events were determined by prior events, from the Big Bang on. For this view, the Big Bang is invested with the natural equivalent of a divine initiating role.

Many modern scientists, however, have emphasized physical indeterminism. The leaders of the indeterministic, random, or chance view are eminent scientists such as Manfred Eigen, Francis Crick, Jacques Monod, and Ilya Prigogine, all of whom have received much attention. The deterministic group is equally illustrious, but its members are largely from an earlier era. They include Galileo, Newton, Einstein, and, as recently as 1972, the chemist of the atomic bond, Linus Pauling. Edward O. Wilson has elegantly presented the case for genetic determinism in the human realm, which places him among the living determinists with Pauling and Pauling's student, Emile Zuckerkandl.

The Institute for Creation Research sees fit to agree with extreme determinism. It quotes selectively the writings of Newton and Kepler, who tried "to think God's thoughts after Him." But these scientists lived centuries before anyone knew that protein molecules are composed of strings of varied amino acid molecules, and even more years before a model for a primordial cell was constructed in the laboratory.

It is to the credit of Galileo, Newton, Einstein, and others that they recognized physical determinism, given the state of science in their time. Now we can cite as a tangible basis for both physical and biological determinism the myriad of specific interactions of organic molecules within living things, and before living things. As constructionism has begun to show, this is a scientific, or natural, source of highly deterministic direction in evolution.

As we have seen, biochemists are, of all scientists, most likely to be selectively attracted to the internal problems of evolution and, in that group, to appreciate the appropriate subproblems. A biochemical orientation is not enough, however. A better understanding of constructionism, which has not been indigenous to biochemistry, is also a pressing requirement.

The psychological problems posed by constructionism—that is, the concept of synthetic retracing of the pathway of evolution—were brought home to me when a famous biochemist and I presented separate seminar lectures at Rockefeller University on the same day. Rockefeller University has an attractive recreation room for guest speakers. After I talked, I repaired to it to test my skill at billiards. There I found my famous biochemist colleague. He had evidently attended my seminar, for he immediately said, as he removed a cue from the rack, "What are you trying to do, play God?" His attitude was undoubtedly an example of "It's okay to analyze what God hath already wrought, but not to shadow his miracle doing." Of course, I was not trying to play God; I was trying to play evolution. An ironic sequel to this incident is that, after this biochemist's death, of cancer, a leading origin-of-life researcher attempted to use the favorite enzyme of the deceased to construct a theory of evolution. The attempt was still a failure ten years later, because it employed modern enzyme systems rather than retracing steps to the first enzymes. The incident at Rockefeller University illustrates again that opposition comes not only from fundamentalists and students who reconcile their early upbringing with courses in science by using figurative interpretations but also from other scientists who are under the effect of subliminal thinking rooted in the long age of miracles.

Beyond the power of science? The reproduction in the laboratory of the event of instant creation *is* beyond the power of science as long as creationists, strict neo-Darwinists, and nonevolution-minded chemists are allowed to define the problem. When the scientific perception rephrased the problem into a stepwise evolutionary context, its solution was seen not to be beyond the power of science.

The scientist can present to the fundamentalist propositions and questions that are the inverse of the statements about problems "beyond the power of science." Does the fundamentalist believe that living systems are composed of cells? Has he ever looked through an ordinary microscope? Does he believe in molecules, which he cannot see, and does he believe in molecules of protein? The fact is that when the Book of Genesis was written (and since then) molecules had not yet been mentioned, protein had not been recognized as a molecular combination of varied amino acids, cells had not been visualized, and the instrumental means for observing cells, the microscope, had not been invented. Is it science that lacks the power to answer these questions, or is it the creationist paradigm?

Are we even being fair to the pentateuchal authors who gave to us a bright and picturesque answer to a universal question by using all the resources available to them at the time? Can we say of Ph.D.s in biochemistry, hydrology, and other fields that did not exist in 1600 A.D. that they are using the scholarly resources available to them?

CHAPTER 3

Getting Started

To make a modern organism in one step in the laboratory is a grand and hallowed ideal. We have seen why such thinking is an artifact from our mythologically-dominated past rather than a realistic scientific goal. A reasonable hope, however, was first to make a protocell step by step, and then to guide that protocell to evolve into a modern cell, also step by step—much as it must have happened in the real geological world. This production of a first evolvable protocell might then have been close to a grand synthesis. The idea that organisms might evolve stepwise in a stepwise-changing environment is second nature to biologists. It is, however, disturbing to many chemists. The reason for this unacceptability is not entirely clear. Chemists are all educated in what is known as closed-system thermodynamics. Although open-system thermodynamics has come to the fore in recent years, thanks to Nobelists Lars Onsager and Ilya Prigogine and others, tradition is on the side of the older, closed-system thermodynamics. In closed systems, for example, in flasks, the opportunity for significant change in internal environment is sharply limited. Chemists are thus conditioned to think of stable environments. The shortcomings of the laboratory for representing the emergence of life require careful evaluation.

The original objective in our laboratory was to make protein, not to make an organism. The interest in protein existed because all known life is protein centered. Biochemists and physiologists know this. Without being more than dimly aware that this initial research might work out to yield an organism by steps, we began what we now recognize as stepwise research.

When this research was started, other relevant ideas were just beginning to surface in bioscience or were still underground, awaiting sunlight for the intellectual sprouts to push through. One breakthrough was the identification of DNA's role in inheritance. Many scientists on the fringes of life science in the 1950s and 1960s, and some within that area, focused on this subject. Once it was understood that the large DNA molecule in living systems could *be duplicated,* the idea that this molecular copying is the basis for inheritance generated widespread excitement. A number of illuminating new experiments and many useful concepts resulted, but the initial approach to these complicated processes also led to several kinds of confusion, in the minds of nonbiochemists especially. The relationships between DNA and proteins were often treated simplistically.

One other main development at about this time, however, brought clarification rather than confusion. This was the concept of selforganization, the process that occurs, for example, when protein molecules organize themselves into microscopic organelles. In our laboratory, the idea that a first protein molecule could have organized itself fed on the finding that modern protein molecules tend to organize themselves. This notion was developing at the same time that the broader concept of "selforganization" was being helped by experiments defining primordial protein. As often happens in science, the general idea benefited from a specific application just when a specific application benefited from the general concept.

We now know that proteins organize into various structures: into membranes, into RNA-proteins in ribosomes (the tiny factories of protein in the cell), and, in the beginning, into the first cells themselves. The research began with the objective of making protein. The interest in protein was understandable. A true student of life's chemistry cannot learn the structures, processes, principles, and phenomena of life without encountering proteins at every turn. Proteins are essential to life itself, in both general and specific relationships, and to the process of inheritance. Furthermore, as Robert Shapiro has pointed out in his 1986 book *Origins,* the biochemists' stress on protein has tended to be obscured by the emphasis of nonbiochemists like Francis Crick on the overall significance of DNA.[1] An example of the failure to see the protein trees because of the DNA forest is the title of Crick's 1981 book—*Life Itself.*[2] A better choice might have been *Inheritance Itself.*

Three decades after the essence of inheritance was identified at the molecular level, we can place this great advance in perspective. By 1970, Crick had found that a number of disciples had read into this finding ideas that were not truly there. He pointed out, for instance, that the "Central Dogma" of information flow of DNA→RNA→ protein could not necessarily be applied to the origin of life or to the origin of the genetic code. More broadly, we can now say that the whole of the specificity was not necessarily originally in DNA but in some supercontrol involving both DNA and protein. This larger control is still only incompletely understood. Crick's 1981 book does deal with protein, but merely as a secondary kind of material, not as the primary stuff of life. Life itself results in fact from an orchestration of many substances: water, proteins, vitamins, lipids in membranes, nucleic acids for inheritance, and others. In essentially equating life with nucleic acids, Crick has not provided a model for the emergence of life on Earth. Instead, he has hypothesized life's origin on some unnamed planet, partly because he cannot show how DNA could have come into being without other DNA.

When we sort things out, we see that life had to come first and inheritance later. The notion that inheritance came first and life later has not made sense to a number of biochemists. Moreover, the information built into the stuff of life, protein, has been explained by experiments that show how informed protein arose first. It is thus no wonder that the evolutionary sequence that would put nucleic acids first has for thirty-five years had virtually no experimental backing and that Crick has felt a need to consign the original mechanism to distant space and to bring it here by directed spaceship.

What a Way to Make Protein!

The de novo origin of informed protein has also needed explanation. That both protein molecules and nucleic acid molecules are large distinguishes them from other molecules in living systems, except polysaccharides. The protein molecules are also set apart from most molecules in the living system by the extent of their variegation. Both nucleic acids and proteins are linked assemblies of smaller molecules. In DNA or RNA, the smaller molecular units are nucleotides. In proteins, these units are amino acids. The number of *types* of nucleotide in modern RNA or deoxynucleotide in DNA is usually four. The number of types of amino acid in modern proteins is typically twenty.

This great potential for versatility collectively in large part explains why protein molecules have served as a molecular mainstay in evolution. Any one protein molecule is already somewhat versatile because it contains twenty different kinds of amino acid, which means at least twenty different kinds of reactivity. The biochemist calls these locales in the protein molecule *reactive sites.*

In addition, large molecules like proteins can and do fold in and over themselves. Reactive sites can and do act on other reactive sites to modify one another and to produce new reactivities. Looking at the resultant structural possibilities and their shadings, we see that the potential variety of functional activities is not just greater than twenty. It is, in fact, enormous.

No other molecules in nature begin to provide even a tiny fraction of the functional possibilities inherent in protein molecules. Their physical properties, appropriate both to evolution and to biochemical events, make them the only compounds in nature that can provide the vast array of functions that modern life uses. Because of their flexibility, protein molecules can do many mechanochemical tricks like stretching in muscles. Nucleic acids are effective banks of information precisely because they are less flexible.

It is to the arrays of protein molecules that we must look for the many varying physiological traits and behavioral characteristics that we observe in man and other animals. In some cases, small structural differences in these molecules can underlie large behavioral or physiological differences.

The early 1950s brought the first demonstration of this property of proteins, by Vernon Ingram and his associates. Ingram examined hemoglobin, a protein molecule responsible for the red color of blood. Hemoglobin consists of 576 amino acids in chains and carries oxygen through the bloodstream to cells throughout the body. The significance of hemoglobin to life is such that a leading protein chemist in Munich, Gerhard Braunitzer, has devoted more than twenty years to analyzing hemoglobin from numerous species.

Using methods worked out for assigning sequences of amino acids in proteins in the 1950s (methods based on work initiated in our laboratory in the 1940s), Ingram found that one change of one amino acid in a molecule containing 576 amino acid residues was all that was needed to convert a normal, disk-shaped red blood cell into a sickle-shaped cell responsible for the disease sickle-cell anemia. This difference can be shown schematically as follows:

Normal hemoglobin (576 amino acid residues in toto)
A segment:
Valyl-histidyl-leucyl-threonyl-prolyl-*glutamyl*-glutamyl-lysyl-

Sickle-cell hemoglobin (576 amino acid residues in toto)
A segment:
Valyl-histidyl-leucyl-threonyl-prolyl-*valyl*-glutamyl-lysyl-

The abnormal hemoglobin molecule responsible for sickle-cell anemia has only one amino acid different from the normal. Valyl replaces glutamyl in just one position in chains of 576 amino acids, all the rest of which are otherwise the same.[3]

The idea that the color of one's skin or eyes or the size of one's ears could be traced to the amino acid sequences in protein molecules had won widespread acceptance by about 1960. But psychological behavior is often abstract and harder to sink one's experimental teeth into.

Claims of identification of peptides supporting behavior have appeared since Georges Ungar in Houston claimed in the 1960s to have identified a protein that caused dark avoidance in animals. Such results are difficult to repeat, as Ungar himself stated. He also printed eye-catching words on pioneering scientific subjects in general. Ungar comments:

> Skepticism is one of the basic virtues of the scientist when it is allied with an open mind. It takes great skill to navigate between the Scylla of uncritical credulity and the Charybdis of dogmatic rejection of everything that does not fit readily into the framework of established concepts. Few new ideas have gained recognition without resistance, and controversies are part of the dialectical process by which science advances.[4]

Some of the peptides to which specific behavioral characteristics have been attributed may act by raising the general tone of the organism. In this picture, with uncertainty about what has been learned and with much more detailed knowledge yet to come, materialistically inclined persons are nevertheless intrigued by increasing study of the relationship between matter and mind.

In modern life, an extremely complex DNA and RNA molecular apparatus is needed to arrange the necessary molecular structures. But the exquisite array of manifested functions is, as the examples suggest, often or always rooted in protein molecules, and indirectly in the DNA and RNA that figure in their inheritance.

In the early 1950s, when our laboratory started research on the problem of how protein originated, all the work by others on sickle-cell anemia and on memory molecules was in the future. We knew only that proteins were the basis for all manner of specificities; we had been sensitized to this possibility by a 1940s remark of T. H.

Morgan—"Fox, all the important problems of biology are problems of proteins." Even in 1950, however, it was clear that the modern mechanism for synthesis of protein is extremely complex. How could the first proteins have gotten started in a simple way?

The chemist, whether in industry or in academe, tends to think of the economical process of heating reactants to make something new. Why not heat the α-amino acid components of protein to combine them to form proteins? The answer to this question was generally that this was a hopeless approach. We almost all knew that when we heated a typical α-amino acid component of protein we would get a tar and other unwanted material.

The α-amino acids have the amino group, NH_2, close to the acid group, COOH. Heating such an amino acid could cause decomposition, which, when carried to an extreme, yields black, gooey tars.

$$H_3C\text{–}H_2C\text{–}H_2C\text{–}H_2C\text{–}CHNH_2\text{–}COOH$$

An α-amino acid

In ω-amino acids, the amino group is far removed from the acid group.

$$NH_2CH_2\text{–}H_2C\text{–}H_2C\text{–}H_2C\text{–}H_2C\text{–}COOH$$

An ω-amino acid

The ω-amino acid is not found in protein.

In fact, the du Pont chemist Wallace Carothers thought of heat when he wanted to make, in the 1930s, a substitute for the protein silk fibroin, the essence of silk. Knowing that heating the α-amino acids would give unwanted products according to the conventional wisdom, Carothers heated the other kind of amino acid, an ω-amino acid. This was successful in its own way and led into the whole family of silk substitutes, the nylons.

However, of the twenty kinds of α-amino acid, one provided some hope of producing protein-like molecules by heating. Heating is a not unrealistic assumption about the primitive Earth, because even today's Earth has on it thousands of regions where the temperature for such an event is as favorable as in a laboratory. Using the heat available in the laboratory, two German chemists, H. Schiff and E. Schaal, had each reported about 1900 that they could heat aspartic acid, one of the α-amino acids, and get many molecules of aspartic acid to join together to give what is called a polypeptide, much like a junior protein.[5] Again, the necessary heat is available on Earth even today, near hot springs or on the surface of a Sun-baked desert. This joining together of many small molecules of one kind is known as polymerization. The small molecules are generically monomers; the large ones, polymers. In the laboratory the polymerization of aspartic acid can be speeded up by the use of higher temperatures. But time can be traded for temperature, and there was no hurry on the primitive Earth. Duane L. Rohlfing, a pioneer in the subject matter, found that two weeks of milder heating work just as well as four hours by an impatient chemist, but whether the product of heating aspartic acid was a true polypeptide had been questioned, and properly so.

Eventually, the polyaspartic acid was shown to be a close relative of a true peptide (actually a polyimide). Studies by John Kovacs at St. John's University in Jamaica, New York,[6] and by Allen Vegotsky in our laboratory established the peptide nature of polyaspartic acid, as well as special chemical aspects.[7] By 1960, the early German work was pretty well forgotten as an isolated exception to the rule that one could not usefully heat amino acids. In more technical terms, it was agreed that one could polymerize aspartic acid by heating it but that heating other α-amino acids, such as phenylalanine or leucine, would yield nothing of value.

The key advance occurred through what is known as *co*polymerization. Aspartic acid alone can be polymerized by heat; it is not normally decomposed by heating. By contrast, the amino acid phenylalanine is decomposed by heating. If one heats aspartic acid and phenylalanine together, they copolymerize. Heated together, they give peptides that contain both amino acids.

The central reaction of *co*polymerization is a master key. Amino acids can do together what they cannot do separately. The principle of cooperativity is one that we observe at higher levels, in humans and other animals. Many life forms can do together what is impossible for them alone. The unique value of human and other animal cooperation has its basis in the molecules of which we have been composed since before the beginning of ancient cellular evolution.

How powerful is this principle in regard to amino acid molecules? Our laboratory learned systematically that aspartic acid could be copolymerized by gentle heat with each of the common α-amino acids of protein, one by one. The reaction even looked primitive. Could it be used to combine all of the twenty common types of amino acid into a protein molecule like that unique and supremely versatile polymer of nature—protein? In other words, could aspartic acid be combined with *all* of the α-amino acids simultaneously as well as with *each* of them, one by one? If it could, we might have something like protein in a way so unexpectedly simple that its spontaneous occurrence on primitive Earth could be imagined without strain.

The test of this possibility had its dramatic moments. At the time the key experiment was performed, our laboratory group consisted of the graduate student Allen Vegotsky, the special graduate student Kaoru Harada (who was earning his Ph.D. away from home, Japanese style at the time), and the technician Donna Keith. It is customary to ask technicians to perform the "far-out" experiments. More advanced laboratory workers know "too much" and are less likely to be sympathetic to doing what they may well believe is too unlikely. The experiment designed to test the hypothesis that all twenty common amino acids could be copolymerized in one reaction was therefore assigned to our technician, Keith. When Harada realized that this seemingly hopeless experiment was going to be done anyway, he offered to check the result analytically. He did the analysis by breaking the product back down to its constituent amino acids, separating them on a strip of filter paper (a process called chromatography), and staining the paper to identify the amino acid spots. Vegotsky, Harada, and I made guesses about how many spots (amino acids) would appear in their usual, slow way after the staining.

Harada, the most cautious member of the group, guessed four spots, Vegotsky guessed eight, and I, as principal investigator, hedged my oral bet by predicting two to twenty, with an expectation of twelve spots. Actually, this was the most optimistic bet as well as the most weaselly. Harada sprayed the test paper strip with the color-producing agent and, because he was busy with another experiment, came in and out of the laboratory in the Conradi Building of Florida State University. During the color development, Vegotsky and I stayed put in fascination. Watching the development was a little like watching a horse-race, aided by our oral bets. A few spots appeared quickly, and then nearly all of the rest of them emerged in an unsteady stream. There were finally fifteen or sixteen spots. Vegotsky, Harada, and I looked at each other and said almost nothing.

Although we started with twenty amino acids, two of them, asparagine and glutamine, break down to another two, aspartic acid and glutamic acid, during the recovery for analysis. Another amino acid, tryptophan, was, as expected, totally destroyed during the analysis. So we saw fifteen spots when we had a right to expect seventeen at the most (see fig. 3.1). Minor discrepancies could be explained technically by the fact that some spots huddle together. An expert chemist like Harada could separate these fifteen amino acids by this simple analysis quite cleanly.

This and later experiments demonstrated that it was possible simultaneously to combine, in a ridiculously simple way, all α-amino acids. The key was and is to include aspartic acid or glutamic acid, or both, in the mixture to obtain *co*polymerization.

This simple leap to a kind of protein could easily be visualized as having occurred in the geological realm. Organisms like bacteria were not needed for this trick of combination. The experiment showed the great power of thermal *co*polymerization. On the other hand, the experiment did not allow us to infer that the first protein-like polymer on Earth contained all of the amino acid types found in modern proteins.

The experiments did permit anyone to infer that functionally versatile peptides probably arose very simply. Anything this close to protein was close to a doorway between chemical evolution and biological evolution simply because protein is the stuff of life as we know it.

The simplicity of the thermal copolymerization of α-amino acids

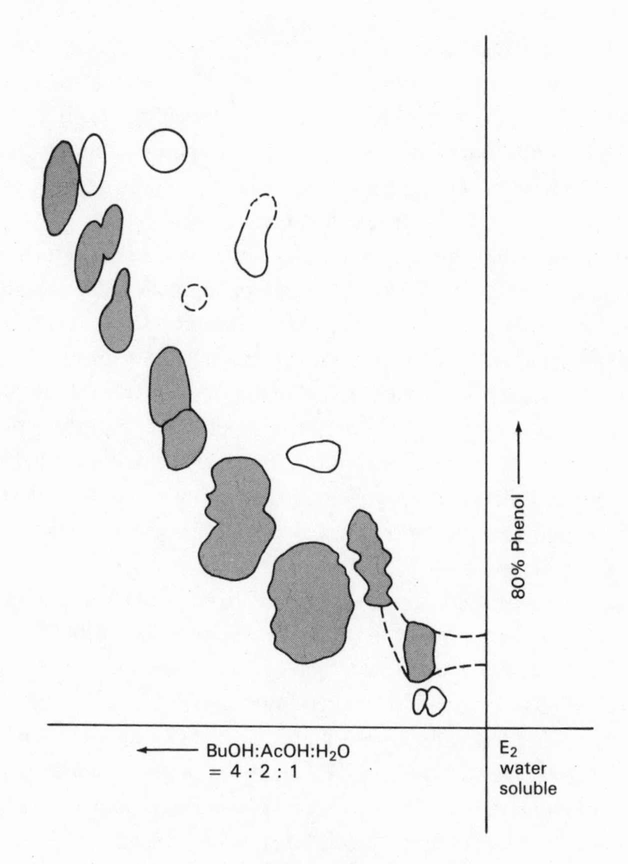

Figure 3.1

Chromatogram of hydrolyzed sample of polymer of the common amino acids.

was made vivid for me after we published our results. I received a letter from a friend, Art Knight, of the University of California at Berkeley. He wrote that he had been told that we had reported a simple way to make something like protein. The product, he had heard, looked like a breakdown of the protein serum albumin on chromatography. He wanted to check the experiment. In response, as soon as the usual reprints of the article arrived from the journal *Science,* [8] I sent one to Art.

Two weeks later, a second letter arrived. "Thanks for the reprint," it said, "but where are the directions?" I then realized that the directions are so simple and brief that they can be, and were, easily overlooked. A simple method of preparation was especially hard to believe for something that behaves like a complex protein molecule. Since that episode, thousands of high school students have repeated the experiment, many of them for science fairs.

Knight repeated the key experiment after his second inquiry and then accepted the result readily. Some of my colleagues elsewhere were also properly skeptical of the original report. However, the head of the Department of Chemistry at Florida State had already handed our written directions to his own laboratory associate at my request. The associate successfully repeated the experiment from the written directions alone, before we submitted the paper for publication. Even so, several of my colleagues objected to our claims and to our referring to these polymers as proteinoids. How could one so easily make something like protein?

One colleague told one of my graduate students that he and a few other members of the department were doubtful of such work, and warned the student that he might get hurt in the process of thesis review. Undeterred by such dire predictions, the student proceeded with his work and later staunchly defended his strong thesis. He received his Ph.D. degree quite as rapidly as did his contemporaries engaged in more routine studies and went on to a productive academic career. Meanwhile, because of such experiences, I became further convinced of the value of having an uninvolved colleague repeat salient processes and results before submitting something new for publication. As I saw over the years the eagerness with which opponents attacked experiments related to genesis, I was glad I had initiated this policy early. While it is often true that opinions are strongest where evidence is weakest, the converse, that opinions are weakest where evidence is strongest, fails to hold. Although the thorough checking did not always persuade strong-minded opponents, it did provide a sense of security for all of the authors of each of about ten unique findings.

Getting Started

Second only to the conceptual obstacle that one could not get anything useful by heating mixtures of amino acids was the preconception that, if one could, the product would be hopelessly disordered. Actually, the two are variants of the same preemptive perception.

In modern systems, the order of amino acids in any one kind of protein molecule is exquisitely maintained and inherited. As was already indicated for hemoglobins and other molecules, the behavioral quality is rooted in the exact order of amino acids in any one protein molecule. For any given biological function, the number of molecules of one kind of protein is ordinarily not small; it may be in the thousands in the mass of material in either the organism or a laboratory flask.

This means that there must be a mechanism for stamping out hundreds or thousands of each kind of protein molecule, all with their amino acids in the right places. In modern cells, that is done by the genetic coding mechanism. The process, carried out cooperatively by DNA with RNA and protein, is remarkably accurate and dependable. It is also extremely complex and incompletely understood.

In order to get started with trying to make a primitive protein, one had to deal with the fact that proteins as we know them have this repeated order of amino acids. But for a primitive protein, we could visualize nothing so complex as a genetic coding apparatus to produce repeated order as in a stamp mill. There were, then, two possibilities, provided one could indeed make a kind of primitive protein.

One possibility was that the primitive protein was disorderly. That has in fact been the widespread assumption. In his book *The Origin of Life on the Earth,* Oparin embraced it when he stated, "All that we can expect . . . is organic polymers in the shape of polypeptides . . . having, as yet, no orderly arrangement of amino acid . . . residues adapted to the performance of particular functions."[9] Oparin here expressed, and helped form, a widely held view.

The other possibility was that the first proteins were orderly. If they were, where did they get the information needed to combine the α-amino acids in specific sequences? There could not yet have been a complex genetic coding mechanism: no DNA, no RNA, no proteins. The only logical explanation was that the reactant amino acids themselves carried the instructions for their own order.

Even though this was perceived as logical after the fact, as are many other experimental findings in science, clues did point to it. The clues we had came from earlier studies of reactions of special amino acid molecules to make special peptide molecules in the presence of enzymes. In these experiments, performed in our laboratory in the 1940s, we thought we saw evidence that amino acids were ordering themselves.[10] This suspicion went against the widely held view that the specificities of reaction were in each case due to an "outside agent," the enzyme used for the reaction. Essentially, no thought had been given to the possibility that the reactants might themselves contribute to their own reactions. The whole subject was in fact formally known as the Specificity of Proteolytic Enzymes. When it became worthwhile to think about the reactions of amino acids by heating them, and with no prebiotic enzyme to help order the amino acids, the suggestion of self-ordering proved to be a lucky heresy. The same experiment that showed that one could heat a mixture of amino acids to yield what would later be called thermal proteins revealed, to a considerable degree, that the amino acids order themselves. Compared to a disorderly product, in which no two molecules would be alike, the idea of a product arranged by internal control opened up a whole new ball game.[11]

Against the long-prevalent channeled thinking that order in amino acids was arranged solely by outside molecules such as proteolytic enzymes or by nucleic acids, these freshly obtained and interpreted results suggested that the order was determined from within. This was appropriate for geological simplicity, for biochemical simplicity, and for forward evolutionary flow. It is also one of many results in science that illustrate an observation of the late Albert Szent-Györgyi, discoverer of vitamin C, to the effect that research is to see what everyone else has seen and to think what no one else has thought. In this example, the newness of the thinking was aided by the fact that no one else had seen the result.

Getting Started

The far-flung effect of comments by wise old scientists like Szent-Györgyi was illustrated for me in 1986, while I was on a trip to Beijing, arranged by a wise old Chinese scientist, Bei Shihzhang, who wished to explore connections between our model and his studies in evolution. As on an earlier trip to China, my wife and I visited a number of laboratories in Beijing and Shanghai. Two of them had quotations on the wall in the entrance foyer. One of these was in Chinese; I was told it was by Szent-Györgyi, but no record of the statement appears in my notes. The other, in English, was the one paraphrased above. Here, cited in China, was a statement by someone who was born and raised in Hungary and spent most of his life in the United States.

The amino acid experiment indicated technically that, prior to the beginning of cellular evolution, there could already have existed ordered proteins. Later, we were to find that the specificity of that ordering went far beyond our initial belief; it turned out to be comparable to what is accomplished by the modern genetic apparatus. Such a comparability has been difficult for knowledgeable specialists to believe, despite verifying reports from numerous laboratories.

The presumption that biological evolution began at the other extreme of disordered, random, or chance events has been widespread. It has been a tenet of what is called neo-Darwinism, not always clearly expressed, and therefore not easy to answer. The other source of antagonistic theory is what physicists call game theory. In this treatment, the amino acids in a protein molecule are likened to identical beads on a chain or to identically shaped cards in a playing deck—hence the name *game theory.* The fallacy here lies in likening twenty types of amino acid molecule to twenty beads or to twenty cards. Each kind of amino acid—indeed, each kind of molecule—has instead its own, unique shape; the reactions are at each stage very specific. By 1986, at least ten laboratories, and probably several more, had reported the self-ordering of amino acids in experiments in one form or another.[12]

That thermal protein molecules are produced in a self-limited number of types makes it much easier to rationalize biological functions in protein molecules of one kind. Those few of my colleagues who were outraged by our claims of protein-like molecules did not take kindly to the reports on internal limitations of structural type. Nor did they respond warmly to the subsequent stream of reports from other laboratories that these molecules had many of the functions of modern proteins. That the degree of activity found was much less than exists in most modern proteins was for them a key weakness, because they did not acknowledge the amplifying power of evolutionary processes, in which they did not much believe anyhow.

The first objection was to referring to these molecules as protein-like. The fervor of objection was a little stronger to the word we coined: *proteinoid.* Nevertheless, the authority on chemical terminology, *Chemical Abstracts,* used the word *proteinoid* between 1967 and 1972. When *Chemical Abstracts* came out with its own term, *thermal protein* (chapter 1), they gave even stronger implicit emphasis to the protein-like nature of the thermal polymers.

The proteinoid properties that received the most early attention were the enzyme-like ones. Since the types of modern known enzyme number in the thousands, they have been systematized. Some enzymes catalyze the breakdown of molecules; others, with help from energy-donating compounds like ATP (adenosine triphosphate), do the opposite job of building molecules; still other enzymes are specially involved in oxidation, transfer of amino groups, and so on. Workers in the area have found representatives of each of the major classes of enzymes in proteinoids as well as in modern proteins. At least one study found hormonal activity in several proteinoids.[13]

Accordingly, a kind of protein can be made in the geological environment without cells. So far, only protein has been experimentally shown to be informed at the outset (when made in the absence of DNA/RNA), and to be the only material that could bridge prelife and life. Also, the thermal proteins are ordered and have numerous biological functions. They have a third biologically precious quality; they can organize themselves into cells. We have mentioned this potentiality before now; we will elaborate on it in the next chapter.

CHAPTER 4

The Protocell: How It Looked

So Many Cells on Earth

The common denominator of all life on Earth is the cell. From the smallest to the largest living thing, life is made up of cells. In fact, the smallest unit generally recognized as living is a single cell. Single-cell organisms are by far the most abundant form of life on Earth.

The human brain, which is an organ—an assembly of cells—is composed of ten billion to a hundred billion cells. So, with six billion humans on Earth, the number of human brain cells on Earth is more than a hundred billion billion, or a hundred quintillion, or 10^{20} cells, obviously an extremely large number. But it is minuscule compared with the number of single-cell, or unicellular, organisms on this planet. Most single-cell types are bacteria, which are everywhere. They cover our skin and inhabit our intestines. The most common form, *E. coli,* constitutes much of our stools. Despite this less-than-elegant pedigree, the availability of *E. coli* and its nontoxicity have made its chemistry the most studied of any organism's on Earth.

Cells are not mentioned in the Book of Genesis. That classic was written and rewritten before microscopes were invented, so any criticism of ignorance or omission on the part of the original authors is unfair.

In the modern pantheon of biological morphology, the cell occupies the central position. It is a spheroidal structure; the more primitive it is, the more truly spherical it appears to be. The cellular structures, especially the primitive, unicellular kinds, fall into a relatively narrow range of microscopic size. The protocells produced in the laboratory are quite precisely in this range, as well as being nearly spherical and remarkably uniform in size and shape.

That there are now colossal numbers of unicellular organisms on Earth, that all fit a similar size range and shape, and that the retracement experiments also suggest an original spherule of one to three micrometers in diameter reinforce the interpretation that the first cells on Earth were spheroidal. These specifications are met uniquely by proteinoid microspheres.

Another microscopic structure with a superficial resemblance to cells is the oil droplet. A few of the materials in nature tend to organize themselves into microscopic structures, or droplets, because of the internal attractions of the constituent molecules. We know water droplets, oil droplets, syrup droplets, eye drops, and so on. What looks to the eye like an unbroken stream of water is seen in high-speed cinematography to be a stream of water droplets.

Cells are extremely complex droplets quite different from oil droplets. They do contain oily materials, lipids, in their structure, especially in the membranes, but they are much more stable than oil droplets—primarily because of the protein content of the membrane. The envelopes of these droplets, the membranes, selectively permit molecules to pass through; they tend to favor small molecules for passage.

The modern, evolved membrane consists mainly of an oily substance, that is, a lipid such as lecithin, plus protein. In plant cells, cellulose is the major component. The lipid is a barrier, whereas the protein (or cellulose) is the main source of stability of the cell and serves to form channels that modify the barrier function by permitting selective diffusion of other substances in a way that the lipid barrier alone cannot do. The protein also possesses enzyme activities in the membrane related to those in the protein molecules in the interior, or cytoplasm, of the cell. Moreover, the electrical properties of the cell are rooted in the protein, especially in the membrane. Thanks to the virtues of constructionism (see the next section), we can best study the different functions of protein and lipid ("oil") in constructed models. In addition, these functions are more cleanly segregated in models of primitive cells than in evolved cells, which have in fact been selected in nature to utilize cooperatively the benefits of both kinds of compound.[1]

The proteinoid microspheres are complex rather than simple, they carry out many functions, and they possess, in limited degree, the barrier functions of a droplet of oil, because they are (only) partly oily in their chemical nature. The total range of their subtleties, however, greatly exceeds that of oil droplets, which they resemble most in morphology. The microspheres are easily produced in uniform size, whereas oil droplets are diverse and loose.

Partly because of the ease with which oily substances form droplets *shaped* somewhat like cells, the earliest laboratory models of cells were oily droplets. The contribution of proteins to stability and to the more subtle, nonvisible properties was recognized later. It was not until about 1970 that one pioneer neurophysiologist, David Nachmansohn of Columbia University, put a much needed major emphasis on proteins in cell membranes.

By 1960, only Oparin's coacervate droplets and proteinoid microspheres had received much attention as representations of how the first cells on Earth could have originated. Since then, ten or more synthetic types of cell-like structure have been offered as imitations or improvements. Only the proteinoid microspheres show uniformity of microscopic size and the many biological properties described in this and the next chapter (4 and 5).

Once the idea that life did not have to emerge in its modern entirety—say, as an evolved salamander or as a human—is accepted, the inevitable question is, How did the first cell emerge?

Despite a fair amount of agreement on the definition of the unit of life as the cell, biologists, like other scientists, tend not to agree entirely on any issue. A few dissenters from the cellular definition seem to choose as their definition of life the gene or DNA. For all real purposes, and consistent with all kinds of biological data, the cell continues to be widely recognized as the unit of life. Correspondingly, the protocell is the unit of protolife.

Constructionism

Here it is helpful to consider more closely the value of constructionistic research for retracing the evolutionary steps. When the first NASA bioscience subcommittee took up the question of how life began, I had frequent discussions with a former classmate who was a comember. While the NASA hierarchy wanted us to learn about how life began, and about related questions, he was amused and amazed at our own emphasis on making a cell. He was protective of what he regarded as his realistic outlook. The notion that a living cell could come into existence, or that a scientist could arrange conditions so one could come into existence, bothered him. Since our laboratory had claimed already to have demonstrated that feat in some degree, he often chose to argue the point.

His own experiments, all thorough and quite respectable, were in the field of genetics. He did no experiments on the origin of life, but he did have much to say about the subject from his theoretical perspective. One of his favorite statements was "Maybe we can learn enough in two hundred years to make a cell."

While that statement seems at first to betray pessimism, it actually expressed rank (and empty) optimism, in view of his approach to scientific research. His mode of doing science was analytical, as it has been for the vast majority of bioscientists. We now recognize that one cannot learn to make a cell in two hundred years of analyzing cells. Nor in two thousand years! What was needed instead was to retrace the relevant steps of evolution in a forward and synthetic direction. The understanding of how to do it develops during and from the experiments. The knowledge was found in trial and success, more popularly referred to as trial and error. The necessary approach and philosophy are in the direction opposite to that of analysis; the evolutionary direction has its own principles, processes, and phenomena. One needs the perspective and practice of "Back to the Future" (chapter 1).

By 1960, our laboratory had made a primordial cell, even though at the time we knew only a few of its properties. By 1985, a quarter of a century later, no one had yet made a modern cell, despite attempts by workers like James Danielli to disassemble a modern cell and reassemble the components. Danielli transferred nucleus or cytoplasm from one cell to another.

The necessary acts for our laboratory protocell were the retracement of steps in molecular evolution in a forward direction. Loosely speaking, this is known as synthesis. Strictly speaking, only the preparation of the polymer is synthesis; its assembly into a cellular structure is selforganization. Synthesis followed by selforganization is called construction or cytoconstruction.*

Oparin's approach to the problem was something else. In making coacervate droplets as models for primitive cells, he was practicing selforganization, but not all of construction as defined above, because he did not synthesize his polymers. He obtained them from organisms that were already later products of a lengthy evolutionary sequence. The horse, for instance, had synthesized a precursor of gelatin, collagen, and the acacia tree had synthesized gum arabic.

So, the production of coacervate droplets is a sequence of two processes. The first is analytical preparation from what is already here, courtesy of synthesis by organisms (rather than by the geological realm). The second is assembly.

Because of its history, the coacervate droplet fails to represent the

*About 1976–1978, I discussed the special and somewhat elusive nature of cytoconstruction with our NASA monitor, Richard S. Young. The idea intrigued him. Later he mentioned that he had discussed this approach with others he knew. He reported that one group responded with immediately awakened recognition; another group argued vehemently against it.

first cells, which had to arise from something earlier, not from something later. The making of the droplets does reveal some principles operative in modern cells. Work with them has been pioneering because of its emphasis on organization in general. Their defects as *protocells,* however, are crucial and due mainly to their not coming into existence by a kind of geological synthesis. Because of the way in which their components are obtained, they are simply irrelevant to our problem of how life began. Another major defect that we discussed earlier has had a long (and delaying) impact on the development of origin-of-life theory. That is the degree of instability of the coacervate droplets.

How scientists do their experiments is in many cases a crucial component of their general thinking. The use of coacervate droplets critically influenced Oparin's thinking about the origin of life. His assumption that life had to begin with a cell has received strong support. He was aware of the possibility of organizing cell-like structures experimentally, from the work of the Dutch chemist Bungenberg de Jong, who first described coacervate droplets and the ease of making them. When made of materials from modern organisms, like gelatin and gum arabic, however, they are not stable. They more closely resemble loose oily droplets than rugged cells.

Oparin should have made his laboratory protocells from a primitive material of some kind, but he used modern materials. In using the coacervate droplets as examples of protocells, he showed that he was well aware that he had to overcome the problem of instability. As we saw in chapter 2, Oparin resorted to the principle of natural selection in dealing with this problem.

Oparin's reasoning from his experiments to a general natural picture of the emergence of life left telltale marks on subsequent thinking. The most successful theorists in the subject have been those who pay close attention to whatever experimental knowledge can tell them. The principal pioneers were Oparin, the British physiologist J. B. S. Haldane, and the Harvard biologist George Wald. Wald received a Nobel Prize in 1967 for his contribution to the understanding of the chemistry of vision, a contribution derived from laboratory studies. Wald's superhobby was the theory of the origin of life, on which he published a classic article in the August 1954 issue of *Scientific American.*

Like Oparin and Haldane, Wald saw the need for early proteins, early membranes, and molecules that had the power to organize and even instruct themselves, that is, make choices from among other molecules. It was Wald who recognized that the experiments of a colleague in the Boston area, Francis Schmitt, could be invoked to explain the origin of the first cell. Schmitt had shown that collagen, the main protein in the human body, has the power to organize itself in regular ways.[2] Wald had the interest and perceptivity to propose that the first cell could have originated in this way, that the "right molecules" could organize themselves into cells.

While Oparin made the point of proteins first and brought out the importance of organization in cells, he recognized that the coacervate droplets needed increased stability. Wald essentially repeated Oparin's statements about the precellular need for stability and about the probability that it would be attained by natural selection of precellular types. This outlook persisted for nearly three decades. The experiments and interpretations that changed it involved the making of protein under geological conditions. Thermal protein represents what was present before cells existed, instead of what was on hand hundreds of millions of years later, such as gum arabic and gelatin.

Stability at the Outset

Analytically, the protocells into which thermal protein is converted are very complex, and the processes by which they arise are intricate enough that they are not yet fully understood. Although the product is complex, the actual events that create it are simple. Thermal protein need only come in contact with water to aggregate into cell-like structures. The process is also pervasive. Hundreds of kinds of thermal protein have been tested. Nearly all of them yield cell-like structures.

Some conditions are more conducive than others. For example, if the thermal protein is first dissolved in warm or hot water and allowed to cool, and if the water contains salt (as did a primitive ocean), beautifully formed cellular structures result.

One of the defects of the coacervate droplets noted earlier is their irregularity in size and shape. This shortcoming is consistent with their lack of stability.

Proteinoid microspheres were not produced in any conscious effort to achieve stability in a laboratory protocell. Rather, in the usual manner of research, the microspheres were the result of experiments that started without even the possibility of cell construction in view. Instead, the goal was to make a primitive protein.

When the continuing experiments happened to yield cellular structures, the microscopic units in a sense announced themselves as direct sequels of a primitive precursor protein molecule. Again, this was in contrast to the evolved materials, gum arabic and gelatin, used in the coacervate droplets. The microspheres have stability, or, as one friendly critic stated, they are "robust." There is no need to invoke natural selection to supply stability, as Oparin did. Stability is built into the microsphere from its outset.

Stability has been important in the difference not only between coacervate droplets and proteinoid microspheres but also between one type of microsphere and other types. The first proteinoid microspheres, described about 1960, were originally made of one kind of acidic proteinoid. They are beautiful to look at and have so many cell-like properties that they continue to be used for study of some of their common attributes. However, they dissolve under some conditions, even though stable while undissolved. In order for them not to dissolve, they had to be maintained under mildly acidic conditions in suspension.

Modern cells are in most instances composed of both acidic and nearly basic proteins. By 1962, reports were describing microspheres made of both acidic and more basic thermal proteins. Both kinds of polymers are readily made and are thus easy to visualize as arising spontaneously on the primitive Earth, microspheres composed of both types being somewhat more demanding conceptually, because of need for additional types of amino acid (lysine).

The microspheres that result from the interaction of the acidic and the more basic polymers are stable under a much wider variety of conditions than are the purely acidic ones. They hold together especially over a wide pH range, in many cases in a purely aqueous suspension. Since their first description in 1962, they have been the main type studied.

For the proteinoid microspheres in general, there has been no need to postulate the evolutionary improvement of stability by the selection principle, such as the one proposed for the coacervate droplets.[3] The microspheres reflect the enzymic activities that the chemically basic type of thermal protein possesses: crucial catalytic properties, such as the ability to make peptide bonds and internucleotide bonds, which this type of polymer does with the help of energy-donating phosphates either in solution or in microspheres.

In addition to a robust structural stability and a resistance to the mild alkalinity often imputed to primitive oceans, the microspheres have other sorts of tested stability. They are resistant to moderately high concentrations of salt, to breakdown by water, and to nutritional consumption by microbes from the air (see fig. 4.1). According to some theorists, thermal proteins themselves should decompose back to amino acids in water. They do not do so. The theory behind the question is incomplete; the experimental results alone show it to be inadequate. Undoubtedly, the large thermal protein molecules have internally stabilizing interactions not ordinarily included in such theory.

The experimental facts include a number of subtleties that are not brought out by advanced theorization. For example, the stability and durability of any microsphere varies within that unit. Under the action of water, the boundaries last very much longer. Figure 4.2 is a transmission electron micrograph of a section of a pristine micro-

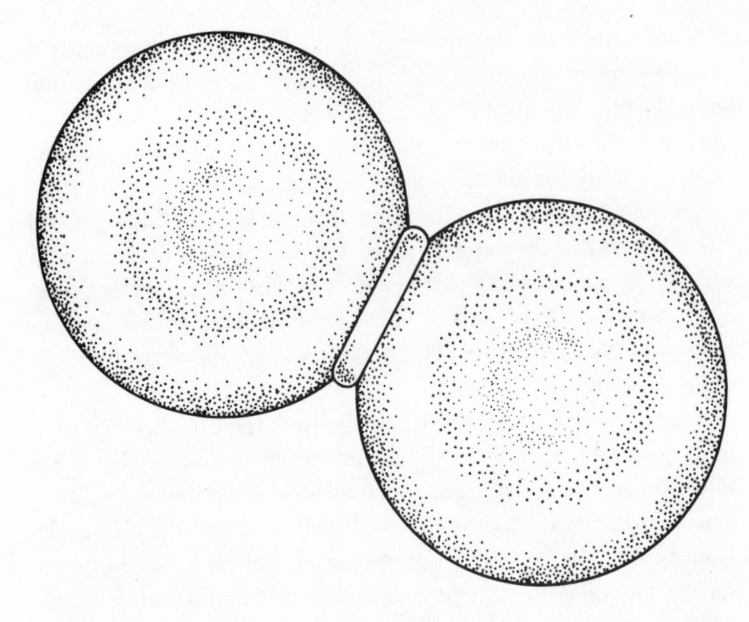

Figure 4.1

The most stable microspheres are those in which acidic thermal protein is reacted with basic thermal protein. They look like the simpler type from acidic proteinoid alone. The complex type holds together under acidic, neutral, and basic conditions. They can be made in the presence of seawater salt, are also stable in pure water, and tend to form couples. The connection results from interaction of the individuals; it is hollow.

sphere. Figure 4.3 shows another microsphere, one that has been leached by water. The contents have almost dissipated. The boundary, a double layer, remains.

In the laboratory, the proteinoid or its assembled microspheres have been found to serve as food for bacteria or fungi from the air. Accordingly, Laura Hsu prepared some microspheres in sterile flasks and maintained them in that condition. They were discarded after six years, their ability to survive as individual cell-like structures having been established. On the primitive Earth before organisms, sterility was the natural state.

The more complex proteinoid microsphere was thus distinguished

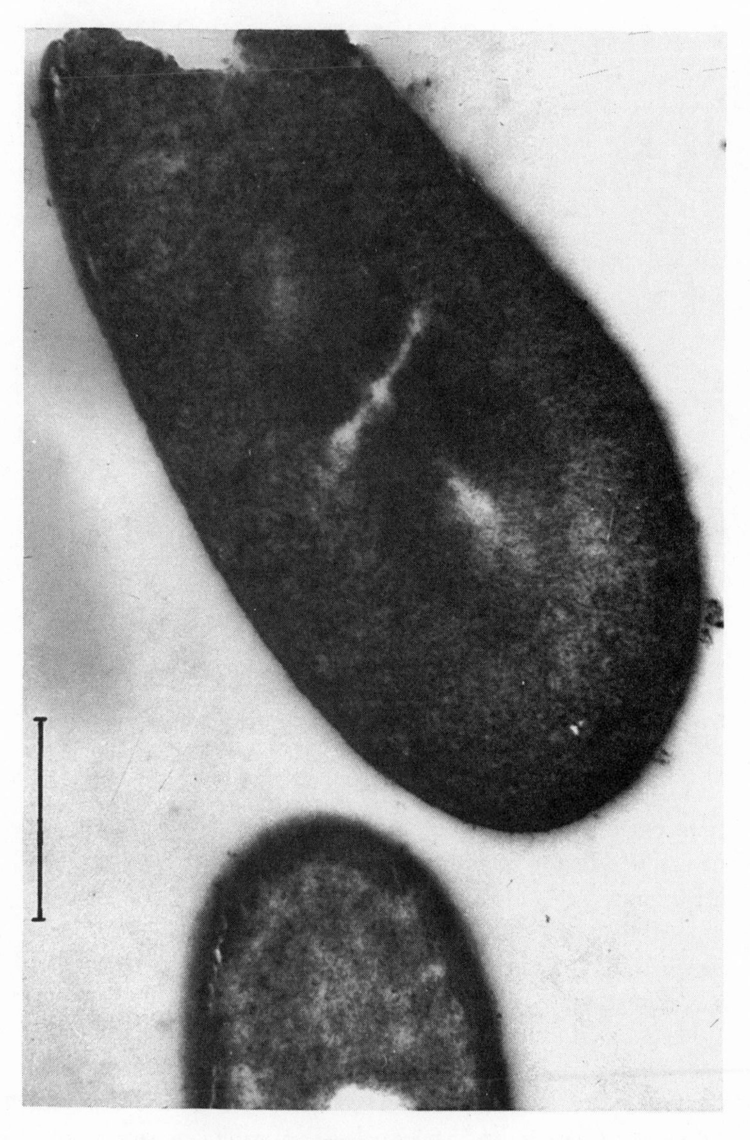

Figure 4.2

Proteinoid microsphere prior to the diffusion of thermal protein from the interior through the boundary.

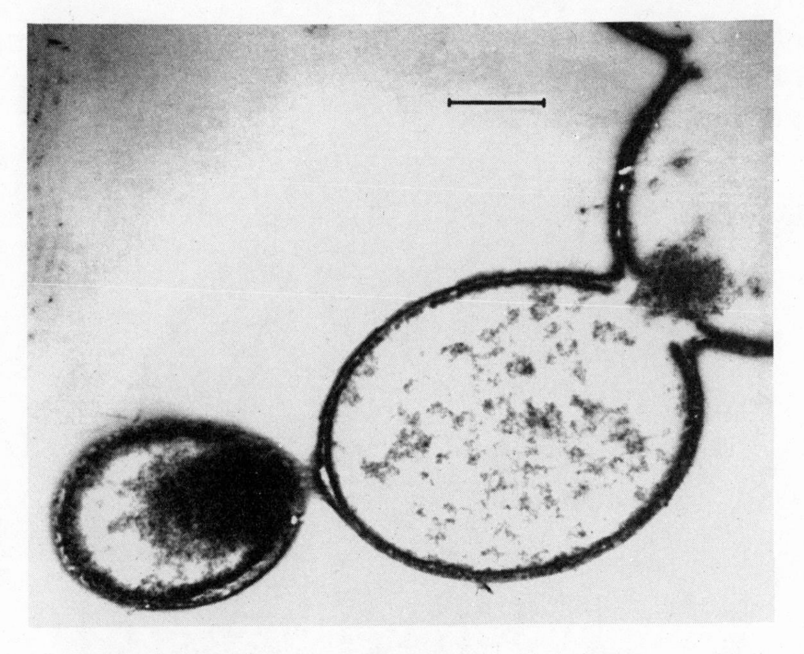

Figure 4.3

Electron micrograph of a section of osmium tetroxide-stained proteinoid microsphere after *p*H has been elevated. Double layer in boundary is prominent.

from the coacervate droplet in that there was no need to invoke natural selection in order for the first structure to evolve to a stable life-beginning cell. Most of the research between 1960 and 1980 was devoted to determining whether the properties of the first cell could be visualized as leading to a modern cell. The answer to the question was affirmative.

Research on evolutionary topics is often shunned because of the uncertainty about what served as a beginning material or assembly. On the other end, the research in evolutionary retracement is practical precisely because one knows what the goal in the laboratory is, since evolution has occurred earlier and since the product is here for comparison.

The Protocell: How It Looked

The cell is a modern entity that has been studied in detail in all its variations more than any other single unit in nature. Accordingly, one could catalog the properties of the protocell and test each function as potentially feeding into that function in a modern cell.

A characteristic of cells of all kinds is that they are all of microscopic size. The largest cells are barely visible to the naked eye. The smallest are about one-tenth of a micrometer, or about four millionths of an inch, in diameter. The smallest cells are believed to be the most primitive.

The proteinoid microspheres conform to this emphasis on smallness for the most primitive cells. The smallest proteinoid microspheres in the optical microscope have a diameter close to that of the smallest cells. They also have a similar, spheroidal shape.

The Lone Ranger

Before the evolutionary emergence of the first cell was retraced, it tended to be discussed as *the cell,* as a lone individual. This was perhaps fitting for a theoretical concept. The reality became clearer when the experiments were done. These made it necessary to conclude that life began not as an individual but as an association of individuals, as perhaps something between a family and a society.

This experimental result answered a longtime question put by behavioral psychologists. In a 1968 paper on social behavior, Ethel Tobach and T. C. Schneirla said, "Until forms intermediate between organic and inorganic matter are better known, the fact that all cells issue directly from other cells may be taken to indicate that no existing form of life is truly solitary—dependence of every individual on others is the prerequisite to social behavior."[4] By the time these authors wrote, the experimental evidence for the truth of their perception that even primordial forms of life are not solitary was already accumulating.

Sociality and Pairing at the Beginning

The experiments showed the emergence was not of a lone individual but of a large group. More than that, the individuals revealed a strong propensity for the formation of couples right from the beginning. The "chemistry," as the popular parlance has it, between two human individuals of opposite gender appears from these experiments to have been ingrained for three billion years.

We can postulate that the attraction is basically between positive charges on one (proto)cell and negative charges on another. Through the microscope, one can see attractions, "flirting," repulsions, and repeated attractions. The experiments have many of the elements of a dating dance or a primitive human ceremony that anthropologists might describe.

The idea of attraction, which is also a yin-yang concept, extends to the sperm and the egg. Biochemical analysis of sperm shows them to be composed of positively charged substances (protamines), whereas the egg has a surface that, like the vast majority of cells on Earth, is negatively charged. Incidentally, that egg surface is consistent with the ancient Chinese assignment of yin, because of negative electrical charge.

But the most exciting suggestion of all this is that the first emergence was both an individual and a social emergence. That the individual is unique, important, and influential is not contradicted by the experiments. It became clear, though, that individuals affect individuals—that societies are groups of individuals acting on individuals. These effects are evident from the outset.

We are never alone.

CHAPTER 5

How the Protocells Behaved

Perceptions of Protocells

What could the first one-cell organisms do? What were their functions?

To ask these questions is to walk the fence between chemistry and biology. The study of the *structures* of the materials making up cells is chemistry. The study of the *functions* of the cell is biology. The study of the functions of the materials within cells is biochemistry.

Chemistry focuses on structures of molecules; biology, on functions of whole organisms. Usually, students are trained almost exclusively in one subject or the other. Only a small number are educated in both, since organization to provide such an education is not widely available or set up very efficiently. The overwhelmingly detailed knowledge that crowds either discipline complicates a thorough education in both.

Biochemistry and *molecular biology* are terms that suggest attention to a zone of knowledge between chemistry and biology. Biochemistry has tended to concentrate on enzymes and metabolism in the life of the organism; molecular biology, on the interrelated functions of the macromolecules.

The unspoken emphases in these subjects led conventional biochemists and conventional molecular biologists for a long time to shun any issue as philosophical as that of evolution. Trained as a biochemist, I was introduced to ideas of evolution in my twenties by the biologist T. H. Morgan. My continued and active interest led a highly honored friend of mine, a biochemist, to ask me in about 1950, "Why do you spend time with evolution? That's something one doesn't do until he is too old to do anything else." He put the question to me when I was not yet forty years of age.

We can now see that the investigation of a protocell required both basic chemistry and basic biology. The study of the biological behavior of an unevolved cell was not feasible until a reasonable model of it was at hand. We needed to know the functional abilities of the whole primordial cell. It became clear that the one way to begin to answer the question was to approach it from the prior side, the purely chemical side of the first cell, and to retrace the steps leading to that cell. That requirement existed because we do not find on the surface of the Earth any primitive cell that we can certify as "primitive," only descendants of primitive cells that we think of as their progenitors.

Associations of Protocells

As we can see in figure 5.1, laboratory cells tend to associate, to form couples and groups. Observing the cells under a microscope, the viewer sees them in constant motion, called Brownian motion, after the Scottish botanist Robert Brown. The conditions necessary for this behavior are many and determine the degree of the activity. Groups can arise immediately as individuals move around in Brownian motion. That the production of laboratory protocells leads not to lone individuals but to groups suggests that life was social right from its beginning.

In order for Brownian motion to occur, the particles must be below a certain, small size. This size restriction is explained by one hypothesis about what provides the power base for Brownian motion—the

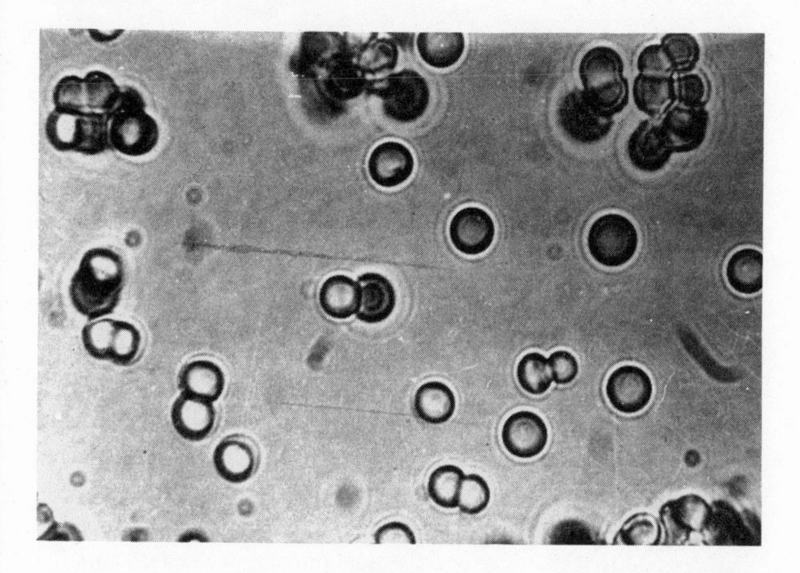

Figure 5.1

The tendency of proteinoid microspheres to form couples and larger associations is especially evident in this photomicrograph.

shoving of visible particles by constantly moving invisible molecules in solution. The microparticles are in suspension, rather than in solution and invisible. The larger the microparticle, the less likely it is to be moved by nudges from molecules in solution. Other things being equal, the smaller the visible microparticle, the greater its motion. The motion also increases as the temperature rises. Brownian motion has attracted the attention of famous physicists, including Albert Einstein, but remains incompletely understood and continues to receive study.

"Dating and Mating"

The phenomenon of Brownian dancing by laboratory protocells has been caught by the camera on a microscope. The activity has the appearance of a dating-mating dance.[1] In figure 5.2, one can see many individuals performing their Brownian jiggling. The four pictures were taken ten seconds apart. As the party progresses, some couples are attracted to each other. A and B are such a couple, as are E and F. C and D are mated throughout. These sequences illustrate only a small fraction of the possibilities. In many cases, the two members of a pair date, separate, date with other members, perhaps repeat the cycle, but finally join permanently in the mating to replace mere dating. Some proponents of both social dating and matrimony have argued that marriage and its usual prelude in Western society have a very long history. If this model of origins is correct, as we believe it to be, the history of matrimony is truly ancient.

It is not difficult to assume that such popular expressions like "the chemistry between two individuals" have a long evolutionary justification. Chemical composition must be responsible for attraction and repulsion, temporary or permanent, between individual cells. We know that when two individual microspheres that approach each other are richly loaded with mainly negative groups (carboxyl) or mainly positive groups (amino), repulsion is favored. Numerous other chemical groups within the molecules in the cells selectively modify these behaviors. The acidic (negative) and basic (positive) groups remain, however, as the primary determinants.

Even though individual cells are either predominantly negative or predominantly positive, each has some chemical groups of other sorts. This means in part that any two individuals can attract each other even though their chemical constitutions are predominantly antagonistic.

Constructing a metaphor in terms of "chemistry" between a human male and female would not be the first instance in which the

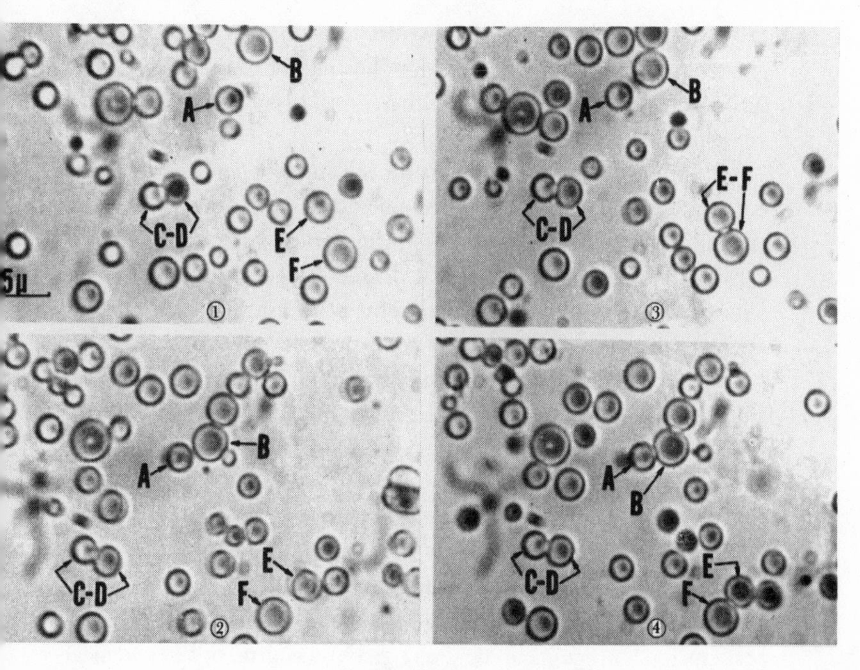

Figure 5.2
Mating Dance

Time-lapse photographs 10 seconds apart. A, B join. C–D are mated throughout. E, F join and remain mated like A–B.

folk perception was adopted before the scientific and evolutionary backup arrived. For the "scientific community," the steps need to be spelled out in enough detail to become acceptable; that consumes much time.

At the human level, the means for attracting has developed to employ more sophisticated chemical devices—for example, sex hormones, male and female. Organismic morphology, too, plays a role—as we all know, even though we may not think in these terms. In human morphology, we find a main basis for visual attraction.

When we extrapolate to the social level, beyond that of the pair, interesting results emerge. In the multitude of interactions observed in the laboratory experiments, the instances of attraction (affection) greatly outnumber those of repulsion (hostility). From this, a moralist can deduce that friendship rather than hostility is a natural evolutionary legacy for mankind.

A huge number of experiments for the study of conjugation (pair-bonding) and the like have led to some other inferences. The typical experiment lasts three days. During this period, Brownian motion is continually observed, the greatest degree of activity coming at the outset. The possibility exists that activity is basically ceaseless, and decreases because of the coherence of particles.

The majority of microspheres have an "adult" life between the first half day and the last half day of the usual three-day experiment. Prior to the first half day, their activity in motion is at its height and appears on the average to correlate with an absence of "dating-mating" tendency. This tendency is more evident among the individuals with more restrained activity, that is, older ones.

Further evidence for aging is found in the last half day, at which stage most of the microspheres appear to be too tired or too old to participate. The aging can be prolonged by a control of the chemical composition. In a population of all highly negative (say, rich in aspartic or glutamic acid) cells, for example, some Brownian motion has been seen to occur for as long as three months instead of the usual three days.

When two microspheres are well aged and form a bond, the bond tends to be fixed. The collar-like bond clearly visible in figure 4.1 is a common feature, a new structure produced by the two partners. This is the result not of the synthesis of new matter but of the sharing of old matter.

Formation of Junctions

Figure 5.3 was produced by Laura Hsu, a microbiologist who joined our group of investigators and subsequently earned her Ph.D. in biology for the work she did. At the time, a number of us had been

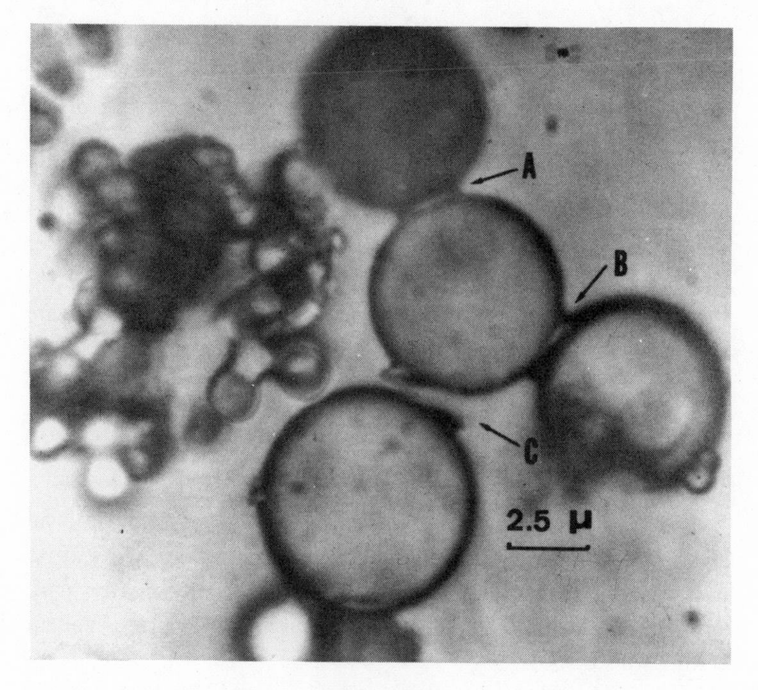

Figure 5.3
Junction Formation

Junctions of proteinoid microspheres. A, intact junction, B, cracked junction, C severed junction.

seeing junction formation for years.[2] Observations at this size level, below one micrometer, through the microscope are necessarily suspect, however. We were not sure whether we were looking at true structures or at optical artifacts, the result of the lens's physical properties when very small particles are examined. Hsu put those doubts to rest. She applied a popular bacteriological stain, Crystal Violet, and obtained the result shown in figure 5.3. It shows four microspheres, with three junctions, A, B, and C. The four microspheres have a very similar size. Often remarked on, the uniformity of size reflects the uniformity of the constituent molecules. Evolved unicellular organisms are also uniform in size. The kinds of algae found in home swimming pools offer an excellent example. In their case, the uniformity of size is maintained by genes. But these laboratory protocells have no genes. The cells and the molecules are uniform, that is, nonrandom, because of the self-ordering power of the amino acids, a determinate factor that we infer made for a controlled evolution before genes came on the scene.

The three junctions were formed when the microspheres touched each other. The junctions are new structures that existed only after the primitive socializing. In this case, the functions are old, by three-day standards, and they have also been stained. So, they are artificially stabilized. Junction A is intact. Junction B is cracked. Junction C has been cracked and has come asunder. One need not be a microscopist or any other kind of expert to perceive that these junctions are material structures; they cannot be optical artifacts.

The aging that we have been examining is aging at a fundamental level. There can be little doubt that the most thorough way to understand and control aging is to learn to do so in such systems. Aging starts in molecules, in cells, in outgrowths of cells, and in connections of cells.

The experiment showing the microspheres' ability to form connections is subject to aging was performed about 1970. Some five years later, a first observation that the ability to conjugate could be revived by adding fresh polymers was tested. In 1986, this reviving effect was tried on modern rat brain neurons by Franz Hefti, an experimental neurobiologist. His attempt to find an effect of simulated primordial polymers on true neurons was surprisingly successful. The significance of such results is that it may be possible to delay the decay of memory. To carry this research out will perhaps require two more five-year periods. In order to delve into it, however, it is necessary to see and honor evolutionary relationships.

The tendency to form cellular structures is shown by virtually all of the polymers made by the heating of amino acids. The tendency to form hollow, filamentous outgrowths is limited to polymers of a more restricted group. These contain high proportions of certain hydrophobic (water-fearing) amino acids such as tryptophan. All the fibrogenic polymers are evidently hydrophobic. All of the thermal polymers have the tendency to form microspherical cells.

What is equally remarkable about these microspheres is their great avidity to conjugate, as was shown, through hollow collars or junctions that they make. This is then a basis for cellular communication, which in modern cells is of two kinds—fast and slow.

The slow one is the more obvious. In the experiments, Brownian motion again catches the eye. First the cells form junctions in a group. One can see them in a limited number of configurations, as in figure 5.4. One places the assembly under a cover glass on the microscope slide (1000×) and draws the water in which they are suspended through them by placing a piece of filter paper at one end of the cover glass.

The microspheres are quite full of polymer at the beginning of the experiment.[3] That state is not shown here. As the water streams through, the polymer in the interior is drawn out through the boundary, or membrane, of each microsphere in the assembly. Other remarkable phenomena then become apparent. Although the polymer in the material is capable of solution in the water and of diffusion through the membrane, the boundaries are durable. They last many, many times longer than the matter in the interior, which diffuses out.

As the polymer diffuses through the boundary, the residual interior particle grows smaller. As it becomes smaller, it becomes jumpier, undoubtedly because the smaller particles in suspension are activated more by the forces of Brownian motion. So, the cells retain their size, the boundaries and the cellular connections last, the endoparticles (residual particles) grow smaller, and the Brownian motion increases. When the endoparticles become smaller than the collar connections between cells, they pass through the junctions at times because they are now smaller than those junctions, which are hollow. Indeed, this passage was the first evidence that the junctions are hollow.

This entire process is precise. The precision can be seen in that the endoparticles in various of the microspheres are all about the same size at about the same time. The process can be stopped at any point.

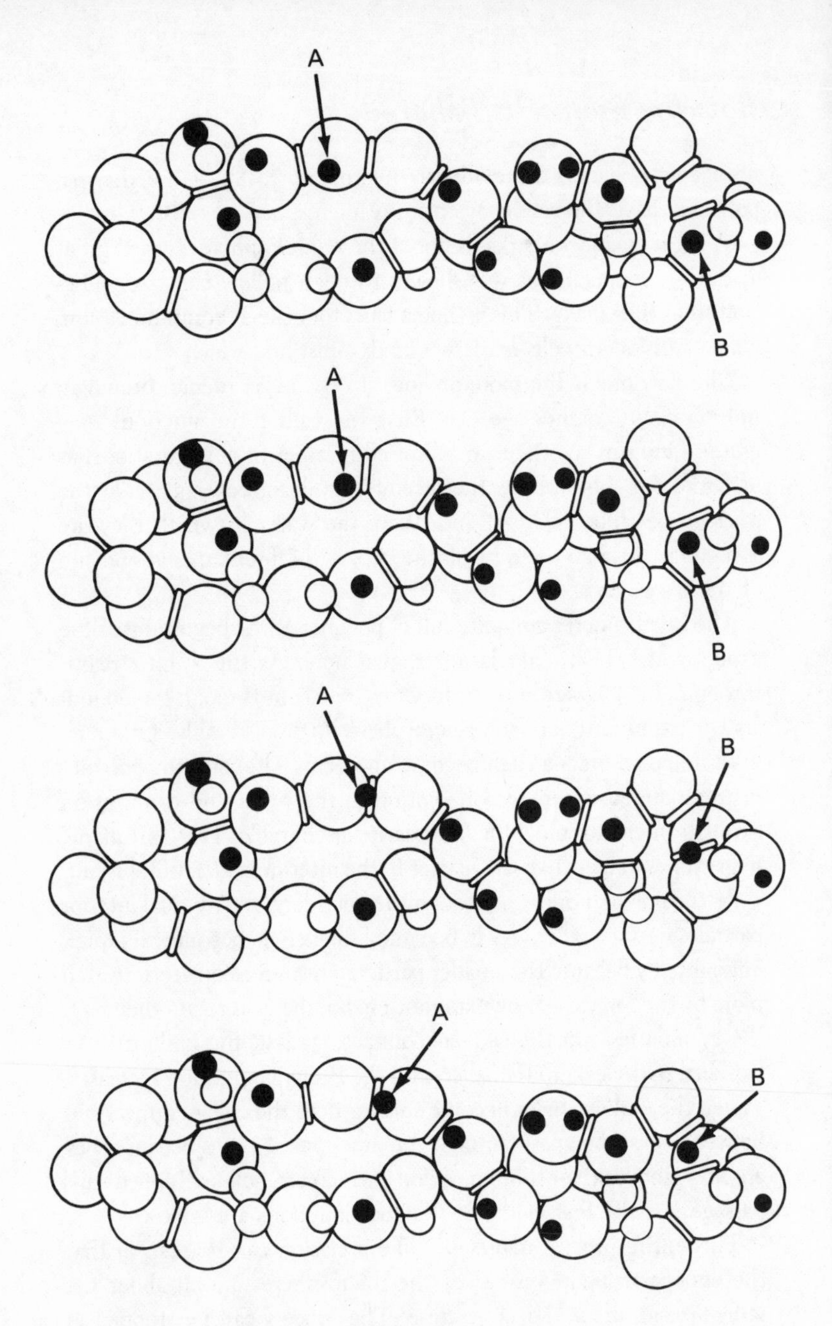

Figure 5.4
Beginnings of Communication

The beginnings of communication at the protocellular level. Transfer of endoparticles (A and B) between microspheres. Each frame is from a time-lapse sequence taken at 10-second intervals. Endoparticles oscillate within confining microspheres until they happen to pass through hollow junctions.

Is This the Beginning of Communication?

Communication between one organism and another begins with the one-way transfer of information. If the endoparticles contain information, they can qualify as primordial bearers of information, and communication can be visualized for them. That would fit in with their structural nature, which is social.

The endoparticles contain information, according to a biochemical definition of the term *information* (a selective interaction of one molecule or system with another). That fact had to be established in other experiments. In these, the proteinoid that composes the endoparticles, or proteinoid much like it, has been found to react specifically with other proteinoids, with polynucleotides (junior RNA or DNA), and with substrates for enzymes. This is information at the biochemical level; the particles contain such information.

The transfer of information-containing endoparticles corresponds to the slow communication now known to exist in modern neuron assemblies. Indeed, Raymond Lasek of Case-Western Reserve University concluded in about 1980 that such transfer of information in modern cells is managed by protein particles. This is much like the process shown almost ten years earlier with the primordial model.

Although not recorded as often or as easily as the transfer of particles, the transfer of particles *back and forth* has frequently been observed. In other words, the information-containing packet can go in either direction, thus qualifying as primitive communication.

The observation that slow protocommunication was inherent in these systems led to the question of whether fast communication might also be intrinsic. In modern cells, fast information transfer is managed by electrical discharge, that is, by streams of electrons, rather than by the slower-moving chemical packets. The question was whether electrical signals, like those common to (communicative) nerve cells, are intrinsic to proteinoid microspheres.

The answer is affirmative; proteinoid microspheres do respond to electric current, generate electric current, and possess electrical potentials. The electrical studies have developed into a whole new catalog of data and interpretation. They will be described separately in chapter 9.

Can the Laboratory Protocells Reproduce?

If cells have the properties that almost everyone looks for, they should be able to reproduce. The ability or tendency to reproduce is high on any biologist's list of the properties of living things. When in evolution did reproduction begin? Can the laboratory protocells reproduce? The answer to the latter question is yes. At least four modes of reproduction have been recorded.

BUDDING CYCLE

The easiest mode of reproduction to perceive and record is that involving budding. Yeasts, bacteria, and hydra are especially involved in this type. The buds form on the parent, then grow and divide further. Alternatively, they separate at an immature stage and grow to the size of the parents.

Figure 5.5 illustrates this process for the artificial cells. The first picture shows an assembly of three proteinoid microspheres seen through the microscope. Again, we see a marked uniformity in their size. So, a sibling-like similarity in size occurs in "offspring" as well as in parents.

Buds appear on the microspheres. In the next frame, buds are seen to have separated. For purposes of monitoring what happens to them, they are stained with the vital stain Crystal Violet. The stained buds are then transferred to a warm solution of the same proteinoid from which the microspheres were assembled. In this experiment, the buds are transferred to the nutrient solution at forty degrees Celsius. The solution is then allowed to cool to twenty-five degrees Celsius. Forty

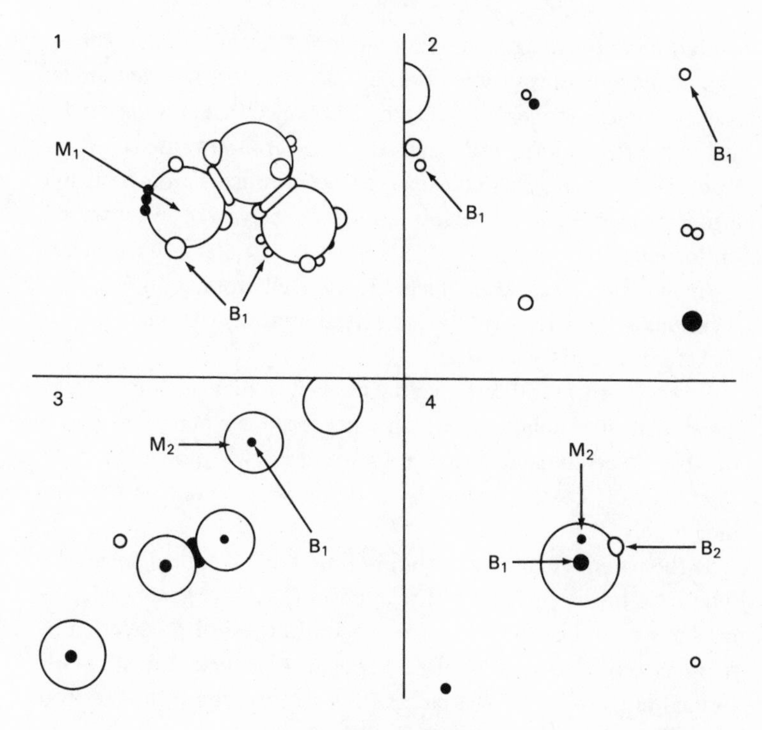

Figure 5.5
Budding Reproduction

Primitive reproduction through budding: 1) M_1 first-generation microspheres; buds have appeared. 2) Buds come off, are stained. 3) Second-generation of microspheres grows by accretion, as explained in the text. 4) Bud B_2, appears on second-generation microsphere.

to twenty-five degrees is often a diurnal temperature range even today.

During the cooling, fresh proteinoid separates from solution onto the stained bud, causing it to grow to the size of the parent, when it stops (if the parent was made at twenty-five degrees). This is growth by accretion, or heterotrophic growth—meaning growth from the outside.[4]

At microscopic dimensions, one can see stained buds that act as centers of a new generation of microspheres. Again, this new generation of individuals is very uniform in size. This fact has confounded some micropaleontologists, who merely assumed that only once-living cells can be so uniform. For many years, they arbitrarily used uniformity of size to distinguish nonliving cells, now fossilized, from inorganic artifacts. They were not acquainted with first-living, or what we call protoliving, cells.

As in earlier microscopic pictures, the facile formation of hollow junctions is again apparent. But it is the uniformity that most catches the eye of experts.

In the fourth, and last, picture is seen a fully mature second-generation microsphere on which has appeared a bud. This bud is capable of becoming detached and starting a new reproductive cycle. We know of no reason why the cyclic process cannot be repeated indefinitely.

Is the parent-child relationship in the budding cycle a generative type or a foster parent–child relationship? If it is generative, the model is more like the dominant modern process of procreation by mammals. If, alternatively, the bud begins as an actual small, newly formed microsphere that attaches itself to the parent, the model is more like a foster relationship.

In a minor proportion of the cases studied, the bud appears to "push out" from the parent. In most cases, the bud looks like an incipient microsphere that attaches itself to the parent.

Early in the research, a yeast expert, a mycologist, visited our lab for a few days, having been intrigued by photomicrographs he had seen. He handled the budded microspheres and cut off the bud with his scalpel. He pronounced the feeling as identical to that of cutting a bud off a yeast cell. This was further evidence, albeit circumstantial, that cells and their buds are made of the same kind of stuff found in the models.

This was reminiscent of what had occurred some years earlier at Florida State University. A waggish graduate student borrowed a photomicrograph of an assembly of microspheres and took it to Chester S. Nielsen, the internationally known mycologist in the

Department of Biology. This professor was an expert at classifying fungi, to which group the yeasts belong. He was asked, with no prefatory communication, if he could identify "these." Nielsen responded to the question quite seriously. He studied the photograph carefully and then asked the student, "Were these obtained from freshwater or saltwater?" Actually, they were made in saltwater, but in laboratory saltwater.

Whether the buds are produced as offspring or as adherent foster small particles, the parent and the child communicate closely. Professor Walther Stoeckenius, a leading cytologist, showed this by slicing a parent-bud assembly and finding with the electron microscope that the material of the two is in some cases continuous, with no intervening membrane. In other experiments, either the parent or the child-like "bud" was made of radioactive proteinoid or was dyed. This dye or the radioactive proteinoid diffused from its primary site to the other.

This reproduction differs from that of modern budding yeasts, or modern budding bacteria, chiefly in the way the bud obtains its protein substance. A modern yeast bud synthesizes its own protein from amino acids. The microsphere bud obtains it preformed from the environment. The difference, then, is in the modern type's greater independence. Each type has to be fed, however.

Other experiments have yielded a partial answer to the problem of how microspheres began to synthesize their own protein. The vivid fact, however, that all forms must be fed has led to new emphases in understanding reproduction.

We really do not know of any growing reproductive form that is not fed. Humans require food, usually three times a day, raccoons require food, spiders are very effective at collecting food, and so on. Indeed, all attempts at creating automatic, self-reproducing *machines* take little or no account of the biotic processes of nutrition and growth. Conventional confusion on this point is rooted in the assumptive terms *self-reproduction* and *self-replication.* These terms are applied at the cellular and the molecular levels, usually *self-reproduction* for cells and *self-replication* for molecules. Strictly speaking, however, neither term is correct, as the bioengineer W. Ross Ashby emphasized in 1960. The answer to the question, "How does the living organism reproduce itself?" is "It doesn't." No organism reproduces *itself.*

Ashby then explained that there is a matrix, an introduced form, a complex dynamic interaction between the two, and the generation of more forms somewhat like the original one. Cells or molecules *are* reproduced; they do not reproduce themselves. In order for cells or higher organisms to reproduce, they must obtain energy from the environment. This they do by feeding. In many contexts, the distinction between self-reproduction and being reproduced is trivial or semantic. In our context, dealing with how the processes got started, it is crucial.

BINARY FISSION

The best-known form of modern cellular division is that of binary fission, the splitting in two of a single cell (see fig. 5.6). This is found as a model in the primordial retracement, as in the picture labeled "Binary Fission."

Like the buds in the budding reproduction, daughter halves have been fed with proteinoid to grow to the size of parents. This is not as smooth a process in the laboratory as the budding cycle is. This suggests that binary fission has special requirements that were fully met only later in evolution, but that idea is not much more than speculation.

SPORULATION AND PARTURITION

Reproduction in modern organisms occurs in many modes and variations. The number of models of evolutionary precursors of these mechanisms is large, and consistent with the modern scene.

One of these is sporulation. Some organisms develop sacs of spores, which break open and spew out spores, which grow to the size of the parents. It is feasible to make "spore"-rich microspheres and to let them burst to release these bodies.

What looks like an ancient precursor of the mode of reproduction

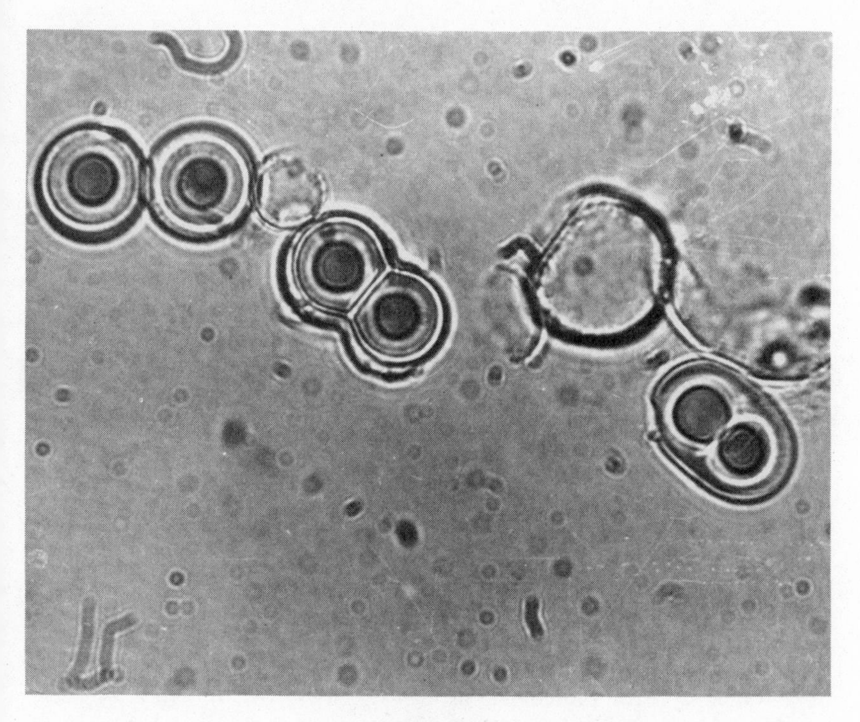

Figure 5.6
Binary Fission

Binary fission of proteinoid microspheres under the microscope. Various stages of what looks like binary fission can be seen under suitable conditions, but the process has not been adequately photographed as a progression.

among mammals, a process of parturition, has also been observed. This process, which bears at least a formal resemblance to the birth of a human body during expulsion of the offspring from the womb, may be the evolutionary precursor. On another hand, it may be merely the behavior of materials and systems for which modern analogues are prominent in our consciousness.

Could Brownian Motion Evolve to Motility?

The protocells are inherently motile, that is, movable. They raise the question of what could have evolved from Brownian motion, the rapid, constant, and evidently haphazard movement of very small particles (0.1 to 1.0 micrometers, or 1/250,000 to 1/25,000 of an inch, in diameter) in aqueous fluids. It is often referred to as random or as apparently random. Aside from its rigorous, statistical definition, the word *random* is often used to mean "haphazard" or to indicate that the observer is mystified by what he sees.

Brownian motion is believed to be due to the propulsive impact that molecules in the water have on particles in suspension, rather than to the intrinsic motion of the suspended particles themselves.

The possibility exists that very small particles of certain compositions move randomly whereas those of others move nonrandomly. The proteinoid microspheres appear to undergo Brownian motion. In the course of photographing several instances of behavior of this type, a few asymmetric proteinoid particles were observed. In figure 5.7, the majority of the microspheres are of uniform small size. In a projection of the film, as in this set of selected frames, one can see that the vast majority of the microparticles are undergoing a Brownian jiggling.

However, one asymmetric particle (indicated by the arrow) consists of two microspheres adhering together, one of average size and one unusually large. It clearly travels around the field of view, while the uniform particles remain almost stationary. Moreover, it rotates first in one direction, then in another.[5] Students in classes who have seen this sequence have argued with the instructor that the particle is alive; it "looks" alive. Some evolutionists like to infer that this movement is an evolutionary precursor to motility in higher organisms.

For figure 5.7, the particles were activated by the addition of zinc and the energy-yielding compound ATP to the suspension. This

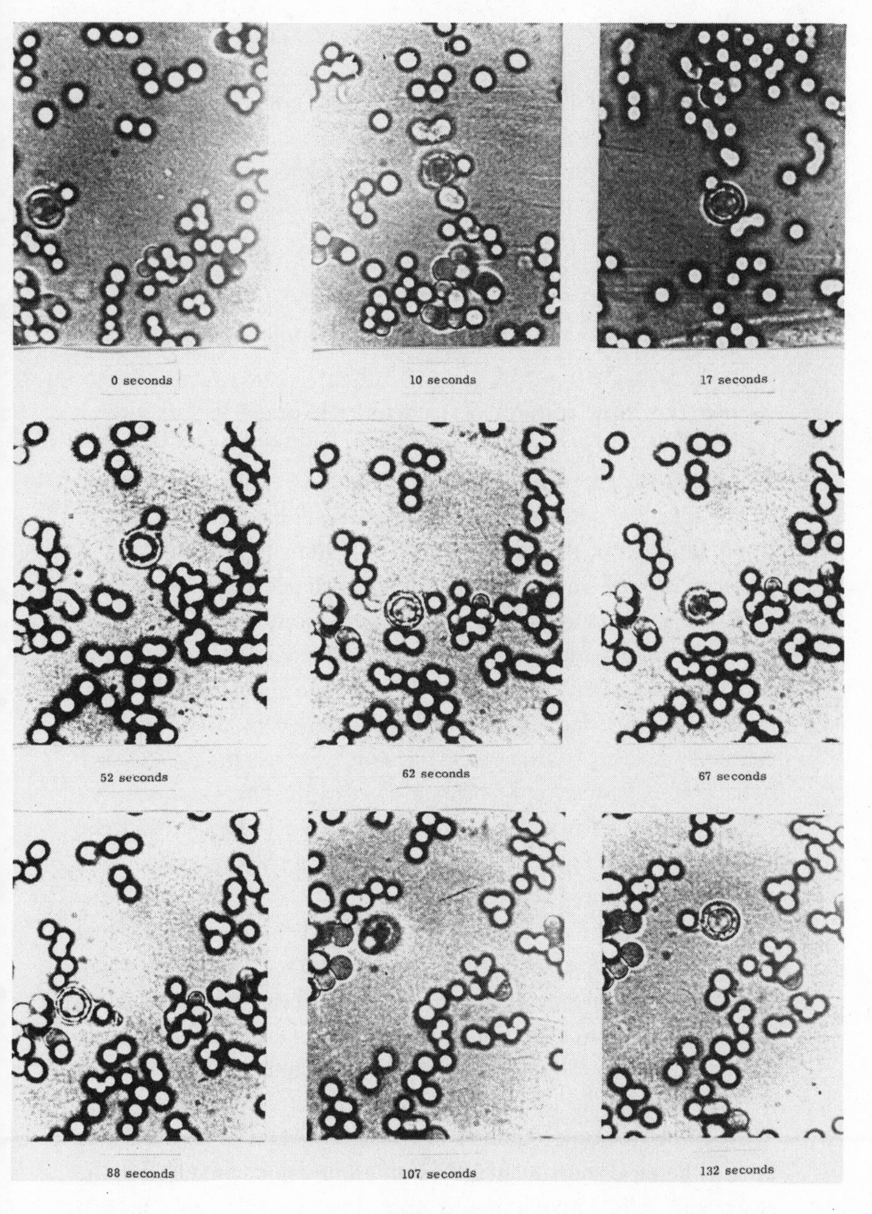

Figure 5.7
Motility

This sequence displays nonrandom motion of one particle. All particles except one are symmetrical.

helps suggest that the observed results were not due simply to the heat of the microscope lamp. Although the same phenomenon has been seen to be less vigorous without these aids, their use probably contributed to the terrestrial evolution of motility.

Metabolic Microspheres

Among the main features of modern cells is metabolism, which comprises the complex enzymic activities that characterize all life.[6] The metabolic reactions are the sum total of cellular chemistry.

The question whether a simulated protocell has metabolic activities reminds us of the famous scientific arguments over *vitalism.* A popular form of vitalism assumes the existence of a special "life force," in addition to physical and chemical forces. This view borders on the unscientific, for scientific thinking is predominantly mechanistic, that is, philosophically causal. The dispute about vitalism versus mechanism in living things was largely settled in the nineteenth century, especially by experiments in which several chemists—Friedrich Wöhler, Justus von Liebig, and Eduard Buchner—showed that the metabolic activity of various cells could be isolated from those cells and that the organic products of cells could be manufactured in the absence of those cells.

For most scientists, this ruled out a special and separate living force. Vitalism persists, however, in attenuated form and subtle variations. The continuing embrace of vitalism has been seen even among origin-of-life researchers. For the beginning of life, the issue is not whether metabolically active agents can be separated from living cells but whether they could have been produced in the absence of cells. This is not the same question as whether metabolic *products* like urea, a chemical constituent of human urine, can be produced in the absence of cells. The question is, Can the chemical agents, protein enzymes possessing such activity, be produced in the absence of cells, or even in the presence of cells themselves produced by synthetic processes?

The answer at the level of both the synthetic enzyme and the artificial cell has been affirmative. Such activities have been generated in the absence of living cells. Some are weak, and some are strong, but all are weaker than those of modern evolved enzymes. Most of the kinds of enzymic activity, the main ones classified by a special commission of the International Union of Biochemistry, have been found in the chemical laboratory preparations.

Origin of Genetic Coding

The enzymic activities observed both in thermal proteins in solution and in proteinoid microspheres in suspension include those needed for the artificial production of offspring microspheres, which would require proteins and, at some later date, polynucleotides (DNA/RNA). In modern cells, proteins are large molecules of weight 10,000 to 200,000 and DNA of much greater size, typically of molecular weight 1 million to 100 million. RNA is typically 50,000. Proteins and nucleic acids each have their characteristic connecting linkage. For proteins, the repeating linkage is the peptide bond, –CONH–; for polynucleotides, the long string of mononucleotides is bonded together by the phosphodiester linkage,

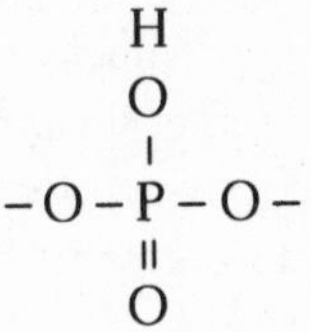

$$\begin{array}{c} \mathrm{H} \\ \mathrm{O} \\ | \\ -\mathrm{O}-\mathrm{P}-\mathrm{O}- \\ \| \\ \mathrm{O} \end{array}$$

A great deal is known about how information is handed from modern nucleic acids to modern proteins but very little about how it entered the nucleic acids in the first place.

The experiments have not shown how large modern polynucleotides could have arisen in the earliest steps of cellular evolution and may or may not have shown how the earliest cells would have made proteins of the size of modern ones. They have shown, however, how the peptide bond and the internucleotide bond would be synthesized. Since peptide bonds make up the repeated backbone structure of proteins, and since the internucleotide bond fills the corresponding role for DNA and RNA, the question of how large the polymers are that can be made in this "primitive" way is the subject of continuing research.

These two syntheses are catalyzed by a basic proteinoid. Proteinoids rich in the basic amino acid lysine have been most studied. It is of special interest that lysine-rich proteinoids, which are themselves basic, catalyze the synthesis of peptide bonds and of phosphodiester bonds alike. This means that one agent can make at least small polynucleotides, as well as junior proteins. This capacity is found in thermal proteins in solution and in suspensions of proteinoid microspheres. Since both reactions are promoted by one agent in one microlocale, we have a candidate for close association in the cell of the kinds of molecule that tend to interact with each other. This kind of control would help the polynucleotides incorporate information that, when read out, could specify the positions of amino acids in protein molecules manufactured at the same time by the same agent in the same microlocale. Thus is visualized, on the basis of experimental findings, the birth of the genetic coding mechanism.

Once produced by the agency of basic proteinoid, polynucleotides react with proteinoids to form salts. It is easy to bring about this reaction with already formed polynucleotides (see fig. 5.8). Research on the internal cellular synthesis of the polynucleotides is continuing. It should clarify the degree to which the protein-like and RNA-like polymers would have had to evolve to become fully modern. The properties of the polymers already formed to date, however, are of a number and kind to explain the beginning of modern protein synthesis by ribosomes and more. Experiments with the simplest kind of RNA model and basic protein model aimed at generating ribosome models have shown that the structures can synthesize peptide bonds and polynucleotide bonds.[7]

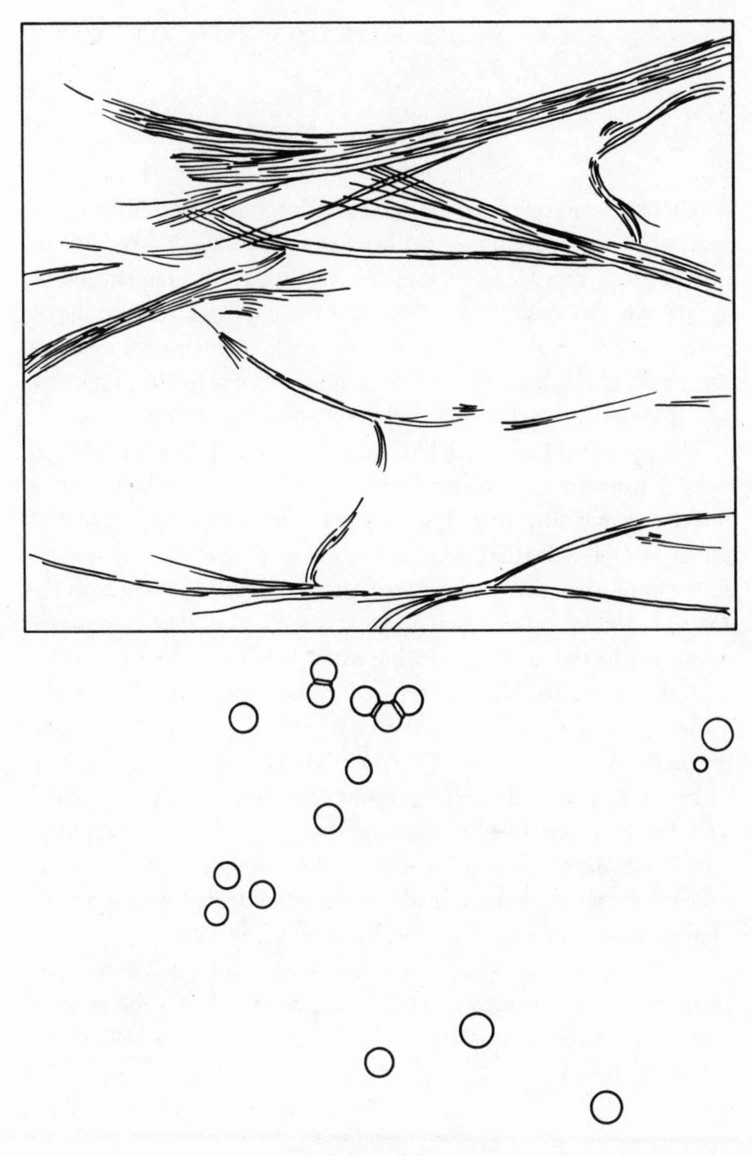

Figure 5.8
Nucleoproteinoid Particles

Interactions of basic proteinoids with DNA or RNA resemble interactions of natural proteins with DNA or RNA. With the former, fibers result; with the latter microspheres resembling globular structures in natural systems result.

Summary

A number of profound thinkers on the subject have long been saying that when an organism is synthesized in the lab, it may not be immediately recognized for what it is. When the proteinoid microsphere was first reported, in 1959, it went unrecognized, except by some able viewers at a distance. The first positive reactions came from organizers of the plenary lectures of the 1959 meeting, in Chicago, of the American Association for the Advancement of Science.

The perception of possibilities by these independent viewers was helpful in assessing progress to 1960, and in stimulating continuation of the research after that. The most thorough critics, including ourselves, wanted nevertheless to catalog the properties of these systems to determine whether they had what could be conceptualized as the roots of modern cellular functions. Major effort and attention were given to this task in the following twenty years.

By 1980, the promise was realized for individual functions in the evolving cell. Some of these required recognizing that the evolutionary precursor was of another type. For example, lipid quality supplied in modern cells by discrete compounds, such as lecithin, was also found to be present in hydrophobic thermal proteins as a result of their composition. The other evolutionary change of major significance appeared to be that in which the self-ordering of amino acids preceded the modern nucleic acid coding mechanism.

The precious quality of evolvability, furthermore, was seen to be the result of the presence in single microsystems of a combination of those individual functions.

CHAPTER 6

From Lab Bench to the Real World

When the focus is on the real world of geological events, the first question should be, Can experiments in the laboratory reliably tell us what happened on the Earth's surface long ago? In chapters 1 through 3, we inferred that the elapse of three and a half billion years need not prevent the drawing of meaningful inferences. We also want to know, Is there any evidence from the field (the geological realm) for processes and products like those employed and found in the laboratory?

For the origin of the first cell on Earth, there is indeed an array of interlocking indirect evidence. We can hardly expect that a multifaceted answer can include very direct evidence of a "smoking gun" quality. However, extremely useful scientific theories have been built entirely on indirect or paradigmatic associations of phenomena and concepts.

The credit for the earliest clear and clean suggestion on where life could begin probably belongs to Joseph Copeland, for many years a professor of botany at the City College of New York. Copeland's essential idea was derived from studies at Yellowstone National Park in the early 1930s, when he was a graduate student at Columbia University. He said little about *how* life began but much about *where* it began. The answer is not simply at Yellowstone but at any place that has conditions for chemical reactions of the kind Yellowstone's hot-springs area makes possible. Literally hundreds of thousands of such places exist now on this planet, over ten thousand in the United States alone.

Before he was thirty, Joe Copeland completed a scholarly treatise on the morphology and nutritional requirements of the blue-green algae of Yellowstone National Park.[1] The algae are in part distinguishable by their green, brown, blue-green, and other pigments. The blue-green algae are undoubtedly the most primitive.

One of the fascinating aspects of the hot-water pools at Yellowstone is their brilliant coloration. Some of these hues are due to minerals, but most are due to the colors of the various algae. The colors show a regular progression, an algal spectrum. Moreover, this spectrum advances or withdraws as the ambient temperature varies, each band of color retaining its relative position.

Another striking aspect of algae and bacteria in the hot pools is that these organisms thrive there even at boiling temperatures. The boiling point of water at the altitude of Yellowstone is about ninety-two degrees Celsius, compared with one hundred degrees Celsius at sea level. At the same time, the hot water runs off into what are known as cold-water pools. Both kinds of pool contain algae, one hot-water algae and the other cold-water algae.

Copeland compared the algae of the cold-water pools that formed in the runoff from the hot-water pools with those in the hot-water pools. He concluded that the cold-water algae were direct descendants of the hot-water algae—that, in other words, the ancestors of the cold-water algae were the hot-water algae.[2] Copeland also looked for the ancestors of the hot-water algae. He found none. These results could be expressed in the following genealogical line:

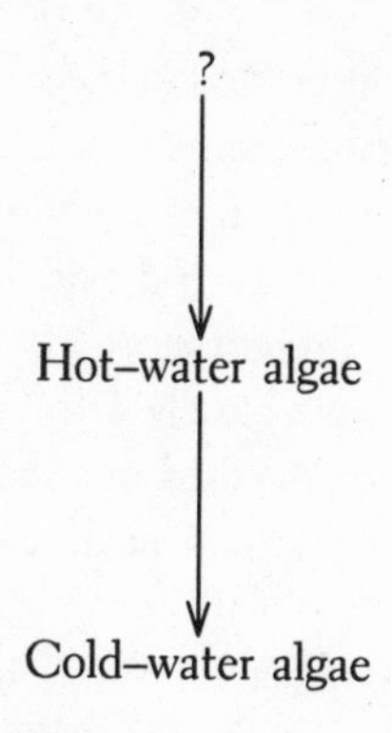

Copeland came to the only conclusion his data permitted: "The morphologically and nutritionally most primitive organisms include the *Myxophyceae,* and their notable incidence in thermal habitats suggests the probability of the origin of living organisms in the thermal waters."[3]

According to this reasoning, life began in the thermal waters as blue-green, spheroidal algae and evolved quickly to other types, as the temperature dropped.

The classic work of Copeland was not widely known; his work came to our attention by a set of fortuitous circumstances.

Two young scientists, Robert and Mavis Middlebrook, both Ph.D.s from Leeds University in England, wrote to me in 1952, at my address at Iowa State College, seeking laboratory positions. Each was at the time employed at Caltech: Bob Middlebrook was a research fellow with Linus Pauling; Mavis had a position with Max Delbrück. Knowing that their appointments would terminate in the early winter of 1952, they had applied for and secured the promise of positions with Albert Szent-Györgyi at Woods Hole, Massachusetts. The one drawback was that the new appointments could not begin until half a year after the old ones at Caltech had ended. So, they asked me about the possibility that one or both might be employed in the interim, at Ames, Iowa, between California and Massachusetts. I was glad to offer them appointments similar to those they were leaving and moving to.

Bob Middlebrook was interested in the same topic that our lab was concerned with—the arrangements of amino acids in proteins relative to their evolution. This kind of study had been going on in our laboratory for over seven years and had inevitably led us to the question of how the first protein arose.

Mavis was content to work closely with me to do the experiments I now was eager to have performed in our laboratory. Bob was free to work independently on amino acid sequence in blood plasma proteins.

On the way from Pasadena to Ames, Iowa, the Middlebrooks' English Ford broke down in Yellowstone Park, leaving them stranded there for a few days in the winter while the local auto mechanic sent for a needed part. During their forced stay at Yellowstone, the Middlebrooks became familiar with the small library at the research station. Its holdings logically focused on reports of studies at Yellowstone park. These included Copeland's report on Yellowstone thermal *Myxophyceae*.

After Mavis began to obtain true peptides by the simple heating of amino acids, she told me about Copeland's paper, which at this early stage in the research provided us all with reassurance. The possibility that a precellular protein was produced by heat seemed compatible with Copeland's findings, although the laboratory results at the time did not yet extend to the production of cells.

Approximately forty-five years after Copeland's work, several groups of investigators proposed that life began in what they called submarine hydrothermal vents—fissures in the floor of the sea from which hot magma in a relatively dry state can escape into a hydrous environment. Early among this group of researchers was John Corliss, who studied submarine vents at the Galápagos Rift, the East Pacific Rise, and other formations.

The hydrothermal vents are much like what lay beneath Copeland's thermal waters at Yellowstone. The conditions at either site are brutal, but experiments show that the laboratory products are rugged. Relatively high temperatures do not readily destroy the kind of organic material involved. The polymerization proceeds, and the self-ordering is unaffected.

From Lab Bench to the Real World

Although the findings of primordial conditions in the Galápagos Rift and elsewhere implied the same conditions as have been used in the laboratory, the suggestion of a submarine volcanic vent was first made twenty-four years earlier. That the prediction and the actual finding were made independently enhances the credibility of both the finding and the idea.

Microfossils

Conditions in the laboratory closely resemble those on today's Earth and therefore probably overlap ancient conditions as well. We have seen that more than one investigative group—studying independently, as far as we know—has identified underwater volcanic vents as providing enabling conditions. The weakest part of these results is the uncertainty whether the organisms now found in the modern vents are truly de novo. An affirmative answer is inferred more easily for the Yellowstone organisms than for those under the ocean.

In addition to similar conditions, investigators can look for products of the earlier times. For evolutionary theory as a whole, fossils have provided the main tangible evidence for the stepwise changes of evolution. For the evolutionist, such fossils are the results of lucky accidents, in which the original structures of an animal or plant were replaced by minerals, to give a product far more durable than the original organism. When fossils are found in strata dating from the most ancient to the most recent, and when the fossils show a stepwise progression in type, we have a fossil record.

One of the more complete examples is that for the horse. Biologists and geologists have joined paleontologists to trace the evolution of the horse. The horse was once about the size of a fox terrier, some sixty million years ago, according to geologists. It nevertheless looked like a horse, and later changes can be traced. The original *Hyracotherium* had four toes on each foot, whereas the modern *Equus* has only one (see fig. 6.1).

This kind of stepwise evolution has been established for other

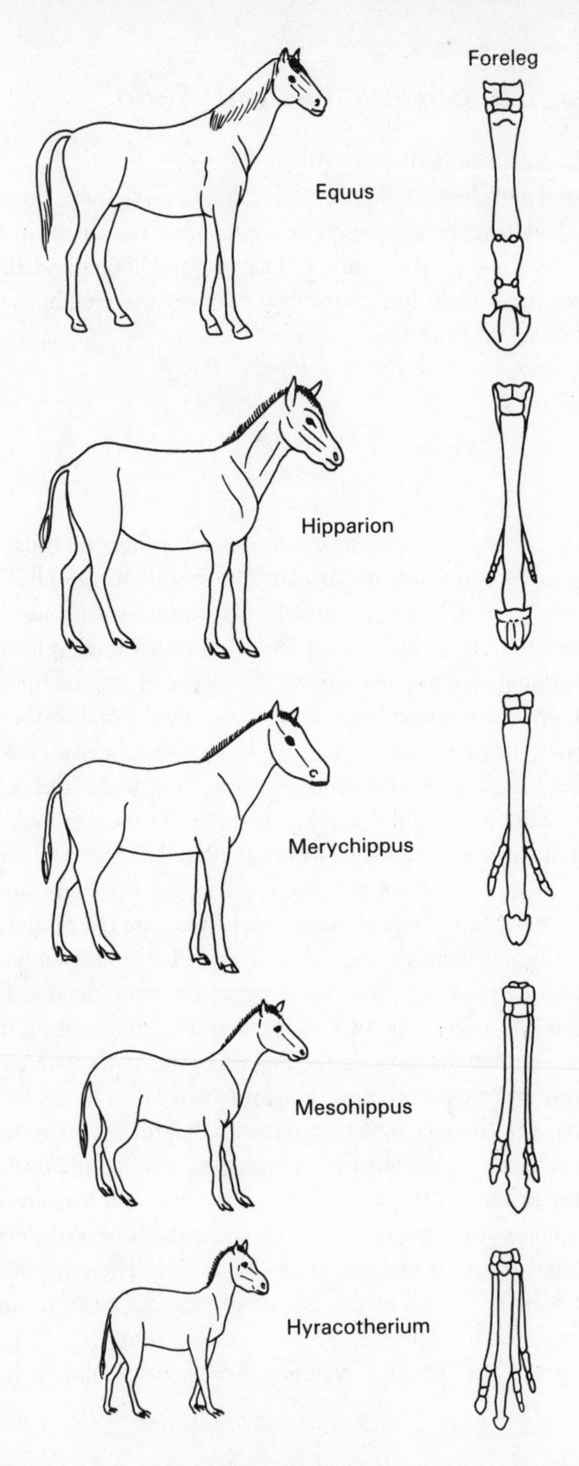

Figure 6.1
Evolution of the Horse

animals: camels, elephants, and giraffes, to name a few. The same approach has been employed, with modifications, for fishes, plants, and so on. With unicellular forms like bacteria, further modifications have been necessary. In studies of microfossils, the microscope has been essential.

A pioneer in this field was the late Elso Barghoorn of Harvard University. His first publication suggesting that, like animals, microbes may have been fossilized and preserved appeared in 1963.[4] Three years earlier, at a meeting in the Shenandoah Valley, he had seen photomicrographs of the then new proteinoid microspheres. Because his microfossils and the proteinoid microspheres from the laboratory looked strikingly similar, Barghoorn was asked if he could be sure that his microfossils were not fossilized microspheres.[5] He did not acknowledge or answer this question until seventeen years later, and then in association with another paleontologist, Lynn Margulis.

When organisms become fossils, they change chemically but little or not at all morphologically. Figure 6.2 compares what Barghoorn and his students obtained from fossiliferous strata with microparticles obtained in the laboratory and photographed before the pictures of the microfossils were published. On the left are nested, vacuolated, and clustered microfossils; on the right are nested, vacuolated, and clustered microspheres (*vacuolated* means that the spheroid contains empty spherical structures). A principal difference is that the microfossils display some of the adherent soil, which is very difficult to remove.

Another difference between the two products is that microfossils are doomed to eternal quiescence. The microspheres can be and have been subjected to numerous tests of activity, the results of which were reported in chapter 5.

A confirmation of the similarity between microspheres and microfossils was reported in 1978 by three workers in the Boston area—Lynn Margulis, Elso Barghoorn, and a student, S. Francis. They were able to carry out the fossilization of microspheres in the laboratory

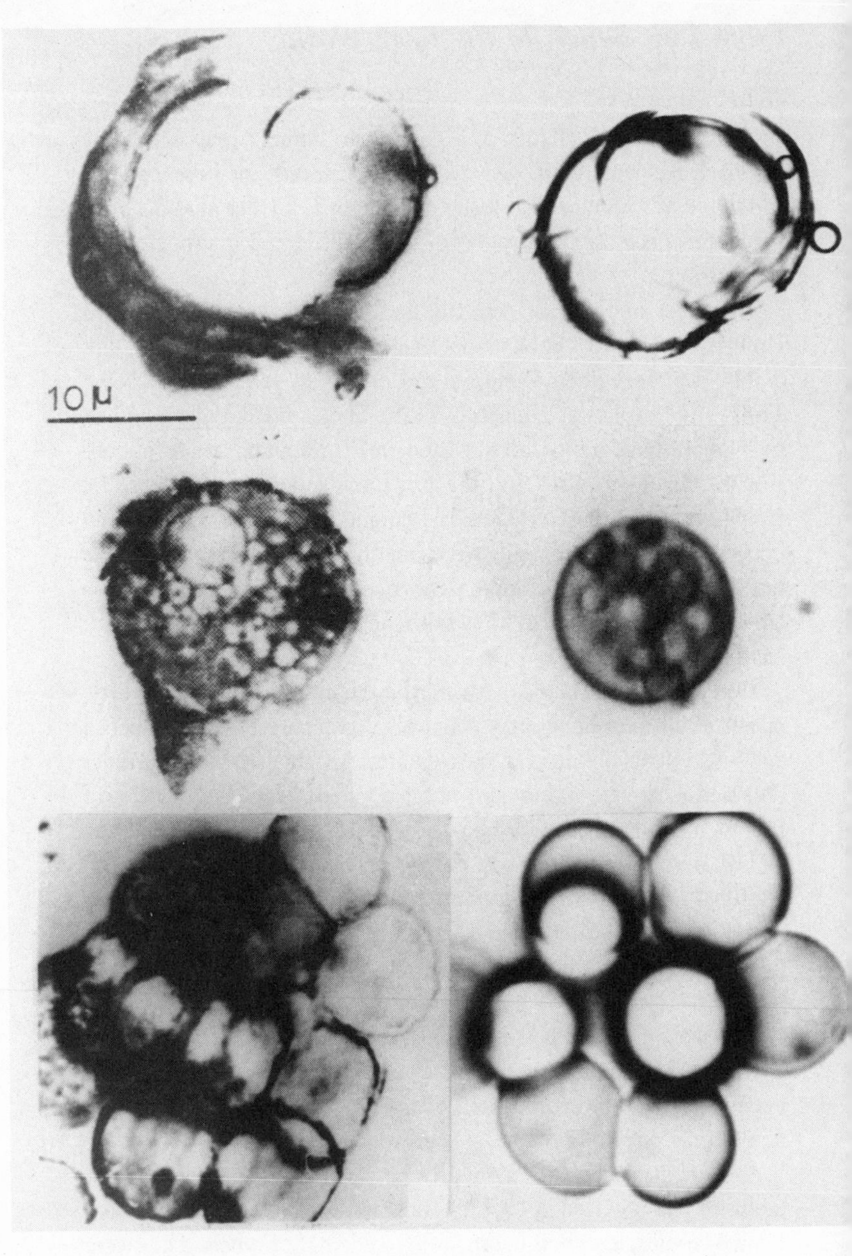

Figure 6.2
Microfossils and Microspheres

Photomicrographs of microfossils of Barghoorn et al. on the left; proteinoid microparticles from the laboratory on the right.

by an ingenious method developed by Richard Leo and Barghoorn. This was Barghoorn's first recorded partial answer to the question put to him seventeen years earlier—whether his fossils might have been proteinoid microspheres.[6]

The main process of fossilization is replacement of organic matter by inorganic silica. Silicon, the most abundant element in the crust of the Earth, can under some conditions replace organic matter in organisms. When it does, a fossil results. The Leo-Barghoorn method of artificial fossilization consists of incubating organisms with ethyl silicate at pH 7, neutrality. Under these conditions the ethyl silicate breaks down to deposit silica in place of the organic matter.

Lynn Margulis, who teaches at Boston University and is famous for her theories on primitive organisms, led the way in applying the method of artificial fossilization to proteinoid microspheres. After her experiments, she wrote that no expert could distinguish the artificial fossil of the microsphere from the microfossils found in ancient strata.

Some micropaleontologists have expressed concern that these comparisons impugn the reality of the microfossils as once-living organisms. Micropaleontologists must always consider that they might be looking at artifacts, such as microspherical assemblages of minerals, that were at no stage in their history living organisms. A much broader interpretation, buttressed by extensive evidence from other experiments, is that the proteinoid microspheres were protoliving units.

The evidence for the occurrence of proteinoid microspheres in ancient strata is circumstantial. However, together with the similarity of conditions as reported by Copeland and by Corliss and colleagues, it greatly strengthens the case for the reality of what is produced in the lab.

Additional support derives from the discovery of the self-ordering of amino acids in modern proteins (see chapter 8).

The recognition that the first organized cells and their material precursors are reflected in the cells and in cellular protein that we see today was not easily attained. The established thinking emphasized the priority of DNA/RNA. That concept was simple but has not been verified. The first molecular information was widely believed to have come from DNA or RNA, both of which certainly play leading roles in the modern transfer of information. While the origin of the DNA or RNA had not been explained, the possibility that informed protein could instead have arisen from amino acids was not widely considered. The newer awareness of the latter arose from the surprising results of experiments done in the somewhat unplanned way in which the primitive planet can be expected to have carried them out.

Since all modern cells contain discrete lipids, the notion that protocells could have arisen without lipids was also not a part of the general wisdom. The natural explanation for that problem was simply that the thermal protein is itself sufficiently lipid-like.

Another difficulty has been the absence of a clear picture of how modern cells arise. No such picture exists yet, thirty years after the demonstration of how protocells could have arisen. Both the first informed cells and the mechanism of synthesis of their precursors have now been shown to have modern equivalents (chapters 5, 8). Proteinaceous protocells containing much information were first made as models. Only subsequently did the collected evidence indicate that the model was correct in its main features: cellularity, contained information, barrier membranes, and especially in the biofunctional properties of informed protein.

Two other consistencies that support the integrated picture need to be discussed. One is biochemical and one biological. Each is the abstract evidence most highly regarded by those with the necessary primary specialized orientation in biochemistry or in biology and also in interdisciplinary orientation. One of these concepts concerns protein as a connecting link between prelife and life, and the other has to do with the centrality of variation in evolution. The first will be discussed here as an addendum to the discussion of necessary conditions and to the microfossil record. The second, which concerns a necessary connecting principle between origins and evolution, will be more fully treated in chapter 8.

The Protein Connection

In science, the pendulums swing far but not necessarily rapidly. The swing between protein and nucleic acid has traced one of the largest arcs. Until about 1950, the primary molecular basis of biological specificity was widely regarded to be in the protein. Then, in 1953, the case was made for nucleic acids. James Watson and Francis Crick demonstrated how the information-laden succession of the nucleotide units in nucleic acids could be copied through DNA. Their demonstration, unquestionably crucial to the understanding of inheritance, is now seen as the copying, in two main stages, of the DNA in one generation for the next generation. The DNA does not copy itself, however; the machinery requires protein enzymes, all of which themselves have specificity.

Few biologists would deny the importance of the role of (protein) enzymes in the biochemical business of the cell, including even DNA synthesis. Biochemists are so familiar with the pervasiveness of enzymes and their actions that they generally see no need to reemphasize it. This view is reflected in a 1980 statement by an expert, Arthur Kornberg of Stanford University, winner of the Nobel Prize for his work on the enzymes that actually make DNA—namely, DNA polymerases. In a book entitled *DNA Replication,* Kornberg wrote, "To the biochemist it is implicit that all biosynthetic and degradative events are catalyzed by enzymes making possible refinements of control and specificity."[7] This means that the replication of DNA uses specific agents—proteins—in addition to parental DNA. In this role, proteins, like DNA and RNA, are informational.

Proteins are thought to serve as the connection between prelife and life because of their various functions and because of what thermal proteins can do. In his 1970 textbook *Biochemistry,* Lehninger listed thirty-nine biological functions of proteins.

For our purposes here, we can classify these functions broadly as structural and enzymic. Proteins are essential to the cell's total structure, to its membrane, and to its enzymic activities, which together constitute metabolism. Other central functions include the biophysical attribute of information content and the physiological one of bioelectricity. The activities most clearly recognized as comprehensively rooted in the proteins are the enzymic ones.

Although these are the principal attributes of proteins in the modern cell, they are also present in thermal proteins. Before experimentation with the thermal proteins, the expectation was that any enzymic activity would be rare and its origin on statistical grounds intricate and very special. The experiments have indicated instead that each thermal protein had an array of *low-level* enzymic activities, weak activities that could have been strengthened by evolutionary escalation. This evolution had a rich seedbed to work on, not something arising from rare beginnings. The Colorado biochemist Thomas Cech discovered in the early 1980s a self-repairing activity of RNA that is catalytic and that has been dubbed enzymic, but it is regarded as rare for RNA molecules.[8] How this might have evolved to an early array of metabolic activities has not been explained by any detailed discipline.

The attributes of bioelectricity and original information have both been suggested by thermal protein, by demonstration. They require a little more explanation. When cellular membranes were first analyzed, the content and role of lipids, fatty material, provided a considerable surprise. As a result, membranes of all kinds were for decades believed by many physiologists to require mainly lipids. From 1970 on, a variety of experiments, but especially the retracement of chemical steps yielding the first cell, showed that protein rather than lipid was the site of electrical behavior. In his 1984 book *In Search of The Physical Basis of Life,* Gilbert Ling, a researcher in Philadelphia, emphasized this contribution: "There is now no doubt that the primary seats of the generation of the potentials are proteins, not phospholipids."[9] That the electrical activity is associated with the membrane protein, rather than the membrane phospholipid, was thus shown by constructionistic experiments to vindicate the perception of the famous professor of physiology at Columbia University, David Nachmansohn.[10]

From Lab Bench to the Real World

The original evolutionary source of biological information has been a problem because of confused perceptions. As we noted earlier, both protein and nucleic acid molecules are informational. It is in the informational context that the chicken-egg problem (see chapter 1) belongs. Both nucleic acids and proteins participate in the modern storage and traffic direction of biological information. This process is extremely complex. We have needed a demonstration of a simple source of original biological information compatible with events on a primitive, lifeless Earth. Here we have learned that the information could have been a complex of instructions from the evolutionary precursors of proteins, amino acid sets, themselves.

The experiments thus indicate that the protocell possessed the first biological information, bioelectricity, its total structure, its membrane structure, and its many enzymic metabolic activities—all because it was composed of proteins. The experiments show that these functions emerged from a realm in which there were not yet any biological individuals as such.

The evidence to date, then, is that protein is the stuff of life, that the first proteins arose in hydrothermal locales and were converted to cells of thermal proteins, and that these protocells had the key properties of modern cells in some degree or in precursor form. The proteins and protocells were the sources of biological information, and the electrical excitability of the protocells was related to that protein makeup. With protein as the material link between prelife and life, and with fossils of the first cells readily available for study, the interplay between field and laboratory evidence has given us a first comprehensive picture of the evolutionary pathway from prelife to life.

Conversation 2

I've been thinking over what you said in our earlier conversation, and I've also read books by authors who claim that a new approach to the problem of life's origin is needed. Do you agree with them?

Yes and no. Yes, older theorists need an approach different from the one they've been using. No, increasing numbers of younger scientists understand the unique value of synthesis and assembly. In addition to the new approach that's already available, experimental retracement of molecular evolution, the older theorists need mostly a new perception.

Why? What's missing from the old perception?

I wouldn't say the problem is what's missing. Rather, the old perception got in the way of the new perception that explained how to design questions and interpret answers in an evolutionary mode. The old notion of instant creation got them into a box, from which they couldn't see clearly into the new realm of scientific thinking. The old perception lingers on subliminally—even for origin-of-life scientists who think *origin* of life.

What is the new perception?

The new perception is through the evolutionary lens.

What are the consequences of the new perception?

The first consequence is a recognition of the need for the constructionistic approach. Almost all, but not all, modern science is analytical and reductionistic. It has yielded a rich harvest of new knowledge but doesn't go far enough. Reduction tears down; construction builds up and moves in the same direction as evolution itself. Secondly, if we want to understand how we came to be what we are, physically and mentally, we have to try to retrace evolution in a forward direction, Back to the Future.

My last biology teacher told me that biology is a science of description, that one studies not principles but classification. Is that still true?

It remains largely true and continues to be an important part of our knowledge. But *combined* synthesis and selforganization (construction) has uncovered principles.

What are the principles?

First, evolution proceeds in steps. Just as one can climb a ladder step by step but cannot jump to the top from the bottom, evolution has made its way in steps. That idea isn't totally new, but it has been perceived only slowly as being applicable here.

Second, large molecules—proteins—*organize* themselves. Chemists have come to appreciate this since the middle of the century. Protein molecules have the reactive groups that guarantee selforganization.

Third, selforganization starts at a stage of self-ordering of the monomeric amino acids that go into protein. This sets the stage for extensive information and variety because the proteins are themselves variegated. The initial result was an extensive yet internally very limited variety of (thermal) proteins. They could assemble and also interact in self-directed ways. From this informed beginning, life arose in steps.

I'll have to think about all that. But it sounds to me like maybe the end of biology.

Not so. Biology is entering a new phase as exemplified by genetic engineering. You can call it *synthetic biology.* It is actually chemical synthesis (structure) integrated with biological evolution (function). The outlook for a larger theory and larger technologies is terrific. Those will require the old biology even more, not less. In education and in applied biology, we'll need a unified, broader view than we've had.

It's clear to me that biologists would not work on the problem of life's origin because it required the skills of a chemist or even of a special kind of chemist. What has held up chemists from working on it?

The answer is sufficiently complex that one has to wonder which aspect of it is most important. The answer cuts across several disciplines; some of it is rooted in the kind of personality that prompted a chemist to become a chemist in the first place. The concept that the right kinds of chemical substance can organize, or assemble, themselves into structures of cellular dimensions extends into undeveloped theoretical subjects outside of standard chemistry. Selforganization, and the attendant specificities, are relatively new ideas. The chemist is conditioned to think that he can make changes step by step with his own hands, but that idea seems to get in the way of the thought that molecules themselves can select and initiate such changes.

CHAPTER 7

Various Places, Various Ideas

We have used and reviewed one approach to understanding the emergence of life. For the sake of balance in this discussion, we need to look at some other theories and clusters of theories that have received attention in the period since 1960. After that, we shall see how the experimental model of the emergence of life alters the essence of Darwinian theory. This development is due to a principle that was derived from experiment and that in late 1986 received unexpected and strong support from the Max Planck Institute of Biochemistry, in Munich.

Another spin-off from the original objective is a first insight into how proteinoid microspheres have become experimental models for learning about the origin of mind. Some quantum physicists have encouraged this newer development since they believe that a theory of the emergence of life must include a theory of the emergence of mind.[1]

Chapter 6 made the point that our original ancestral home was not the modern laboratory. Meaningful conditions in laboratory experiments are, however, essentially the same as those in the terrestrial realm and, perhaps, in special regions of outer space. Some experts assert that the necessary conditions were different in ancient times from what they are now, but the experiments have suggested that the early environment was adequate for life's beginning. When fruitful laboratory conditions prove to resemble current geological ones, instead of requiring theoretical extrapolation backward for identification, we can infer either that we have received a bonus or that many geologists are correct in believing that the range of conditions throughout most of Earth's long history was much the same.

Since Herrera's constructionistic laboratory models, the newer laboratory findings after 1950, and the dissemination of results in 1963, ideas on the emergence of life have proliferated. These will be briefly surveyed in this chapter under headings of locales, cellular functions, and organic informational structures (DNA, RNA, proteins, lipids, and so on). Most attention has been paid to concepts directly or indirectly related to laboratory models and to conditions in the cosmos; some postulates have been omitted because it remains unclear how they fit into our knowledge of the universe.

Could an Aristotelian approach alone have revealed what has been learned by heuristic experiment? We may never know whether we could have arrived at our current knowledge without experiments to lead the way.

In his book on *Origins,* Shapiro shares with Lehninger a skepticism about "armchair speculation." Experiments are also not free of channeled thinking (preface and chapter 1). In what follows, ideas receive some discussion to provide a kind of balance.

Locales

A first question is whether life came from somewhere other than the planet Earth. The idea that it did has been sometimes stated repeatedly, with tongue in cheek. It has not gained lasting support. When viewed solely on an alternative basis in chemical terms, the Earth can claim, even today, a wide variety of natural chemical laboratory conditions. Accordingly, one does not have to assume that life came from elsewhere, if one looks at the question from the perspective of the geographical needs for the first life. Inasmuch as the conditions identified in the laboratory are found on today's Earth, a very strong likelihood exists that suitable conditions prevailed on the earlier Earth also.

OUTER SPACE AND PANSPERMIA

The possibility of life from outer space was promoted in a nontechnical book by Francis Crick in 1981.[2] Crick there expanded an idea he and his colleague Leslie Orgel had first published in 1973. Both the general idea and the specific supporting details had been suggested earlier. An ardent proponent of the view that life originated elsewhere than on Earth was the Swedish physical chemist Svante Arrhenius, a Nobel Prize winner in 1903. For the concept that life is distributed throughout space, Arrhenius adopted the word *panspermia.*

There are two poetic ways in which the claim of a universal distribution of life can be advanced. Even though we recognize that life as we know it in its most elementary organized form is absent, the chemical compounds that produced it are widespread in interstellar organic matter. Second, if mind is in matter and matter is widespread, life is widespread also. In other words, since the decision-making properties of matter are wherever matter is, when those decisions eventuate in life, one can look back and say that the essence of life was in informed matter of the appropriate kind. The precursors of life were thus (a) a simpler form of matter and (b) the universal principle of the self-ordering, or the self-informing, of matter. The two were active together.

For a long time, there has existed a deeply held, but not necessarily widespread, intuitive scientific feeling that life is rooted in all stages of the development of matter. A modern expansion of it is the idea of "The Living Earth." Indeed, the strong effects of medicinal matter, drugs of any kind, support the view that living units are chemical machines. Medicinals function because they are interpolated as chemicals into (bio)chemical activities.

Without the use of such indirect justification as that of prebiotic matter, however, the concept of panspermia applies to the wide cosmic distribution of true seeds of life as we know it. In his modification, Crick spoke of directed panspermia. By this he meant conscious direction of the transfer of seeds of life—for example, spores—from a distant site in outer space to target Earth. A refinement requiring such mobility would seem to be appropriate and restricted to the era of actual interplanetary travel by spaceship, which began in the late 1960s.

Here too, however, the essential idea was not new. Its early promulgation is discussed in Oparin's book *The Origin of Life on the Earth.* There Oparin stated, "It has been suggested that life might have been brought here at some time by the landing of astronauts, that is to say, highly developed conscious beings who could undertake interplanetary journeys. This sort of suggestion is, however, more reminiscent of science fiction than of a serious scientific hypothesis."[3] The counterargument that rockets containing sperms (germs) of life could be sent without internal control of the vehicle by conscious astronauts is, in the present context, a trivial variation.

Oparin continued, "It is interesting to note that, in spite of his ardent belief in the possibility of interplanetary travel, the outstanding Russian scientist and inventor K. Tsiolkovskii nevertheless categorically denied the possibility of this sort of artificial transport of microbes. When he died in 1919 he left a manuscript discussing the possibilities." Oparin further quoted Tsiolkovskii as saying that "this form of transport of life 'with the help of reason' could not have occurred, for no traces have been observed suggesting that at any time or place there have been such highly developed beings deliberately visiting the Earth."[4] What has been learned since 1970 may have made the transfer of germs of life from elsewhere a genuine possibility, provided other difficulties would not have interfered.

As for the charge that his idea is science fiction, Crick reports that other scientists and his wife regard the revived idea as science fiction.[5] Even for science fiction, the statement "There is nothing new under the Sun" is relevant.

However, even if we take seriously the suggestion that life arrived from somewhere else, the basic question remains, How did life emerge? The invocation of mystical distance is unnecessary for those who reason from even approximate biochemical knowledge. Experiments indicate that the necessary informational matrix was right here in ancient thermal proteins—organic bodies shown not to be random and therefore capable of being selected for information. Berthold Förtsch in Munich and Orlin Ivanov from Sofia, Bulgaria, have demonstrated that the source of information, the selfordering of amino acids, applies also to amino acids in modern proteins that are descendants of ancient proteins (chapter 8).[6]

TERRESTRIAL WATERS

Undoubtedly, the most often invoked site for the emergence of life is the waters of Earth, especially the oceans. In 1959, for example, an international meeting on the History of the Oceans that was convened by the American Association for the Advancement of-Science included a session entitled "The Origin of Life."

Another site is what Darwin called "a warm little pond." Principles of physical chemistry, which favor confined locales, make the idea of a warm pond more plausible than that of open oceans. More likely locales are the hot-springs zones suggested by Copeland from his studies at Yellowstone National Park (see chapter 6). In 1981, following earlier proposals, Corliss focused attention on hydrothermal vents beneath the sea (chapter 6). The thinking has shifted to increasingly hot locales.

Essential to the reconstruction of events is the recognition of a change in phase. The making of the precursors can begin in a hot, relatively dry environment if they are ejected into a watery one. This visualization is easy for biologists, who are used to thinking from the outside of organisms. Conversely, some chemists have not tuned into the biologists' idea that, while the inside of the organism was evolving, its environment was also undergoing change. However the process is regarded, the notion of a limited watery locale such as a pond or hot springs can be reconciled with a stepwise emergence of organisms from prebiotic chemical precursors.

THE SOIL OF THE EARTH

The lithosphere—the soil region of Earth—is the preferred site in some theories on the emergence of life. A few years after the discovery of the ease of copolymerizing sets of amino acids in laboratory glassware, a report described this reaction on basalt. It received much international attention, although the conduct of this reaction is to most chemists barely different from carrying it out in glassware in the laboratory. Basalt is juvenile earth, like lava, that is ejected from volcanoes. Its nature is consistent with the theory of an early emergence of the precursor of the first cells.

The lithospheric locale recurred in concepts citing the role of clay, with which the name of A. Graham Cairns-Smith and Hyman Hartman have been closely associated. While either basalt or clay is plausible as a physical support for reactions, Cairns-Smith treats clay as a primary organism evolving to an organism almost as organic in chemical nature as present-day ones are. He suggests a kind of clayey first cell, for instance, in the last of his three books on the subject, *Seven Clues to the Origin of Life,* where he states, "It is certainly no new idea that this most earthy of materials, clay, should have been the stuff of the first life—it is in the Bible."[7]

It is potentially confusing to introduce biblical mythology into what is alleged to be science. For the scientist, the clay story leaves two large unanswered questions. First, just what is a clay organism? Second, how could either clay or organized clay have transferred its information, if it had enough, to the organic cellular structures that constitute our present biota?

The main question, after all, is, How did we get to be here? The surest information we have is on the organic nature of the organisms of which we form a major (and planet-threatening) component.

Various Places, Various Ideas

Structures

Structures are the primary concern of chemists; functions, that of biologists. Our problem involves a blend of basic chemistry and basic biology. Ultimately, we must understand functions in terms of structures and structures in terms of functions. In this section, we systematically consider the principal structures—that is, the main kinds of starting material—proposed in theories of the evolution of organisms.

CLAY AND BASALT

We can begin with the last topic we considered in the preceding section—clay in the lithosphere. As we saw, a key criticism of the "clay idea" is that no one has demonstrated how functional properties (information transfer, metabolism, membrane, reproduction, inheritance, and so on) could have been transferred from a clay to the organic content of a true organism. Indeed, a relevant experiment or a convincing hypothesis that clay, or basalt, has these qualities itself has yet to be demonstrated.*

DNA FIRST

The role of DNA in modern reproduction at the molecular level concerns information and inheritance. The large molecules—DNA, RNA, and proteins—carry the information that can be transferred from generation to generation. Scientists have not agreed on, or even been much concerned with, the question of how information made its way into DNA originally, but their demonstrations of how it is read or fed out have been magnificent. They give us a basic, first explanation for how the molecular information underlying human traits, characteristics, tendencies (to cancer, to musical intelligence), and the like, is handed down to the next generation.

*Basalt, or cooled lava, is as a support in some ways more attractive geologically than clay. The basalt is truly juvenile. It has been tested in experiments in which amino acids have been polymerized.[8] The products were slightly different from those produced in glassware but essentially similar.

The process is one of copying, similar to the copying of a mirror image in another mirror. (Like other analogies, this one introduces certain somewhat unreal simplifications.) The sequence of monomers in the large DNA molecules is transcribed into much smaller RNA molecules and from these translated into protein molecules. DNA, RNA, and protein each contain an independent kind of information. As figure 7.1 shows, the inherited replication is by the DNA. From the DNA, the information proceeds through RNA to protein.

The overall replication process, which is extremely complex and incompletely understood, is in essence a copying of the DNA, as in a mirror. The mirror image is then copied to give a duplicate of the parental DNA. It is especially important to understand the behavior of the DNA, which consists of two strands. This multifaceted process

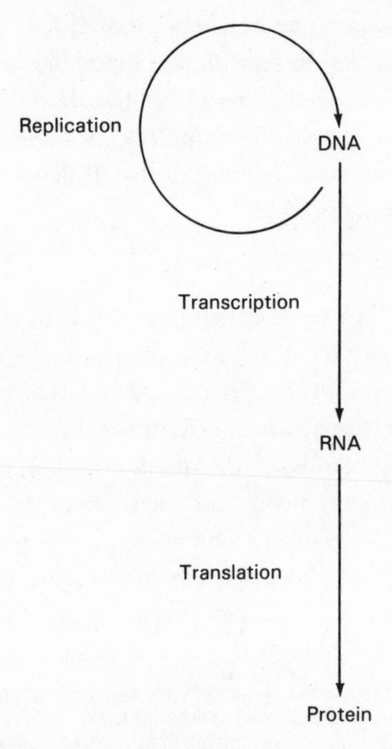

Figure 7.1

must have originated in a long, stepwise evolution. Fortunately, we know that macromolecular information could have arisen by a truly simple process, the selfordering of amino acids. The resultant protein systems could then have aided in the development of the DNA/RNA system as they participated in more modern coding and readout.

In normal reproduction, an exact copy of the original DNA is produced. However, the DNA has not copied itself; it has participated in copying, as carried out by enzymes. The DNA → RNA → protein sequence, the Central Dogma, is a flow of information. It enables us to understand in part that molecules can determine inheritance. As Crick said in 1970, the fact that information flows from DNA to RNA to protein does not explain how original life or the genetic code arose.

The DNA mechanism for the initiation of information transfer has been so influential that alternative modes of thinking are in the shadows. Numerous attempts over a quarter of a century to explain how biological information began with DNA have been widely disappointing. Since a quarter-century of attempts have failed to explain how DNA would have emerged early in evolution, what other theories are there?

RNA FIRST

Attention has shifted from DNA to RNA. By 1981, Crick had acknowledged that the first informational biomacromolecule might have been RNA instead of DNA. Years earlier, enzymes known as reverse transcriptases had been identified. Such enzymes catalyze a flow from RNA to DNA instead of the modern flow of DNA to RNA (again a reversal). The notation is DNA $\xrightarrow{\text{transcription}}$ RNA $\xrightarrow{\text{translation}}$ protein.

The concept of RNA as a first enzyme was aided by the experiments of Thomas Cech, who in 1984 reported that an RNA molecule could transfer one of its own numerous bonds. This suprised those who believed that only proteins could be biological catalysts. It was not so surprising to those who believed that all molecules have information, that all molecules can be catalysts like enzymes. Why was catalytic activity not found in at least one of the many RNAs before 1982?[9]

Another big question about the RNA-first view is how there could have emerged modern metabolism, in which thousands of reactions are catalyzed by thousands of enzymes in the cell. Regarding the flow of information, many have asked, If modern cells follow a menu of DNA → RNA → protein, does this not mean an original sequence of protein → RNA → DNA? We will return to this question later.

LIPIDS FIRST

Probably no event so marked the onset of life as the separation of numerous microscopic cell-like units from their mother liquor, which then became what can properly be called the environment. Various concepts of the first life relate to first structures and first functions of life. They include (1) information, (2) catalytic metabolism, (3) a separating barrier, and (4) reproduction. The third of these has been most widely emphasized by scientists. Once the evolving microunit became separated from its environment by a membrane, so the thinking goes, other evolutionary developments could occur in the interior.

The separating boundary is known to contain mainly phospholipid but it also includes protein. Experiments revealed that thermal protein has most of the properties of phospholipids. Our search for the beginning, however, is bound to be aided by the finding of multiple functions in a single (cellular) structure. This multiplicity is the essence of life itself: not merely information, catalysis, reproduction, individuality, or selective diffusion, but all of these and more—simultaneously.

A popular assumption has been that, like the modern cell membranes, the first ones were composed solely of lipid. The conceptual and experimental alternative to lipid is lipid-like thermal protein. Their similarity is great enough to permit the easy formation of cellular structures containing membranes.

PROTEINS FIRST

In natural selection, speaking crudely, that which survives is that which is survivable. In the step of variation, which precedes selection, evolution makes available the range of functional structures the evolving organism can use. For both origins and evolution, the emphasis has come to fall primarily on natural processes. As T. H. Morgan and later Mae-Wan Ho and Peter Saunders brought out (chapter 8), variation is due to natural physical and chemical processes, and selection is by consequence.

Thanks especially to the research sponsored by the space program, we know that many of the starting compounds for proteins, precursors of the amino acids, have been present abundantly in the cosmos. The making and the combining of amino acids require energy, and we also know that a continuous flow of energy has been available directly ever since the Big Bang. Energy is also indirectly available from the Sun, itself a son of the Big Bang.

Chemically, various amino acids tend to react preferentially with other amino acids because each has two reactive groups: the amino and the carboxyl. Other predisposing chemical features exist in some highly reactive amino acids, such as glutamic acid, a key component of all proteins. As a result, twenty or more kinds of amino acid can combine. When the nucleotides are heated together under conditions that amino acids survive well, the nucleotides decompose. The failure to survive applies to compounds in evolution, much as it does to organisms in evolution. This is not surprising, given that organisms are organized assemblies of compounds. Since organisms are biological, we study them mainly by their functions. The prelude to life is a chemical prelude involving appropriate compounds. Life itself is a biological orchestration involving functions. We are able to analyze the emergence of functions as well as of compounds.

Among those who have independently developed proteins-first overviews are: Klaus Dose, Kaoru Harada, Laura Hsu, John Jungck, James C. Lacey Jr., Koichiro Matsuno, and Duane L. Rohlfing. Matsuno is a theoretical physicist (chapter 8); all the others are biochemical experimenters.[10]

Functions

SEPARATION BY MEMBRANE

When the first organisms were assembled from their (protein) precursors, protoliving units were the result. At that moment, the geological realm became an environment; it had something to form an environment around.

The two prime candidates for the role of separating boundary in the formation of the earliest cells are lipid and protein. Modern lipids are more efficient barrier substances than modern proteins. The kinds of compound that are important to the total economy of the modern cell do not easily penetrate any lipid, either to get in or to get out. Modern proteins do not provide an efficient barrier. We thus do not have at hand truly primitive functional membranes. We can obtain their equivalents—that is, thermal lipids or thermal proteins—only in the laboratory. When we do that, we find membranes of multiple layers. The lipids approximate total barriers, whereas the proteins are a less efficient barrier. As chemical structures, both are lipids—that is, fatty compounds.

Hans Kuhn, author of a number of conceptual critiques in both English and German, has visualized the lower efficiency of membranes as a survival virtue in primitive organisms. As he points out, the kinds of compound needed in archetypal metabolism would have had to enter from the environment through primitive membranes more than is true now. This follows because organisms had not yet learned to make as many chemical compounds as modern organisms now produce.

Although separation from the protoenvironment was the first big step for life, evolution can in fact be considered a continuing effort on the part of the evolving organism to free itself from its environment. This urge is reflected all the way from the first separation to the eventual human colonization of nonterrestrial bodies.

METABOLISM

Metabolism is essentially an integration of chemical reactions in a cell to make more of the same cells. The reactions use energy. We can say that the overall function of metabolism is to convert solar energy to cellular energy by means of those reactions, which number in the thousands in modern organisms. To make more cells, the parent has to make proteins, lipids, cellulose (for plants), and so on. It does this in a continuous energy flow from the Big Bang through our Sun.

REPRODUCTION

Reproduction is often referred to as self-reproduction or self-replication. W. Ross Ashby voiced an important minority opinion in 1960, when he said that nothing reproduces itself (chapter 5). Certainly nothing can reproduce without aid from the environment. We humans, for example, cannot reproduce unless we can remain alive and healthy. A first requirement is energy, in the form of food from the environment. We typically eat three meals a day. We have internalized this to "good eating," restaurants, social gatherings, and the like, but basically we are garnering energy for the continuation of the species. The best excuse that can be made for the use of a wrong word like *self-reproduction* is that the users know better and are employing a kind of shorthand to describe processes for reproduction. Strictly speaking, though, *self-reproduction* is incorrect.

At the molecular level, no one has ever shown that nucleic acids or a protein can reproduce itself, unaided, from its small-molecule precursors. It is possible that, in some indirect way, sets of biomacromolecules can make other sets of biomacromolecules, but this has yet to be thoroughly investigated. Even so, any true process will continue to require an input of energy.

Reproduction in the biotic world is a process of feeding energy through metabolic pathways into the synthesis of "living" matter that assembles itself. We have tended to believe that parents produce their offspring. However, if the processes are sharply self-limited and predetermined, the funneling of energy gives almost the same products each time the organisms are made and assembled. Each production and selforganization of protein is thus not what the word *reproduction* usually connotes, a handing down of metabolic pathways to an offspring that is similar to a parent. Rather, it is truly a reproduction beginning from the latest starting step, but guided in a slightly diversifying way by the parent(s). The energy continues to flow from the Big Bang.

When we question the concept of self-reproduction in the present context of origins, we may come to the variant idea that in each generation the production is begun anew. This is subtly another definition of the word *reproduction,* one that fits the requirements of the bioengineer Ashby and of rigorous reasoning.

INFORMATION

One quality that is often cited as the most needed one to initiate life is information. We have seen that, for the biochemist, information resides in variegated macromolecules like DNA, RNA, and protein. Because of their variegation, controversies have existed over whether the first informational macromolecules were DNA, RNA, or protein.

In experimental retracement of evolution, DNA and RNA are effects, not causes; protein and energy are the causes. DNA is like a recording in a foreign language, and RNA is like a translator. But proteins manufactured the original messages and an untold number of later, related messages. We have as yet just one partial experimental demonstration of how information could have been built into the first protein and how that could have been transferred to DNA and to RNA.

THE BIOCHEMICAL EVOLUTIONARY SYNTHESIS

If we are correct in claiming that key requirements of life are membrane as structure, metabolism and reproduction as processes, and information as control, we may ask, Why not take all of these as an integrated requirement for life? This question has been often posed. We are now in a position to answer it more adequately than before. We will undoubtedly be in a yet better position to do so in the future.

The processes of information generation, membrane formation, reproduction, and (energy) metabolism are all consolidated through protein. Protein or its selforganization products are found to possess all of these qualities simultaneously. This is especially true of the first proteins, which were much less specialized by evolutionary fine-tuning. That interpretation is most easily perceived when we retrace and study early evolutionary steps that made thermal protein and laboratory protocells.

Several Evolutionary Paradigms

In addition to analyzing theories of life's origin by locales, by chemical structure, and by functions, we will find it instructive to review principal clusters of ideas under the names of those who have advanced them. Certain individuals necessarily appear in what follows, whereas others are selected with an eye to attaining a balanced coverage of ideas. Theorists found not to have read their antecedent literature have been omitted.

The choices are also limited to the present era, which has provided a relatively advanced understanding of the material nature of life. The famous Pasteur era in testing of "spontaneous generation" is arbitrarily excluded, because science itself was then too immature. The historical review is available in other books. Oparin's works are most complete in these aspects; his survey in *The Origin of Life on the Earth* begins with ancient beliefs and proceeds through those of the nineteenth century.

HERRERA

Alfonso L. Herrera deserves our special attention because he used an experimental approach to the origin of life and because he had unusual insights. Herrera had to labor to overcome his relative obscurity in Mexico City. His published books, such as *Biologia y Plasmogenia* (Biology and Plasmogeny), reveal him to have been unusually broadminded and highly intuitive. In addition, his experiments were constructionistic. In the process of constructing cells, he made a number of chemically irrelevant structures but also ones that show great insight. Such an array of laboratory products made him vulnerable to criticism by those who looked mainly at the inferior results.

On the positive side, Herrera is credited with having put together simple compounds like ammonia, hydrogen cyanide, and formaldehyde to make polymers and cell-like structures by the thousands. The use of these compounds could have been suggested by what was known about the chemistry of green plants in the early part of this century. It is especially striking that Herrera chose to use the most abundant constituents of interstellar matter—formaldehyde, ammonia, and cyanide—even though interstellar matter was not to be analytically described until thirty years after his death.

Herrera's results would probably have received more favorable attention if Oparin had not chosen to belittle them by likening them to those of Stéphane Leduc, whose synthetic objects (made from inorganic substances) he dismissed. Oparin wronged Herrera by putting his results in the same class. This critique was especially ironic in that Herrera produced his cell models by reacting smaller molecules, whereas for his own coacervate droplets Oparin employed already evolved macromolecules. To assume that life began with materials that had to be products of later evolution is as illogical as to assume that life began with inheritance. For a long time, Oparin failed to see the irony of his criticism.

OPARIN

Popularization of the chemical origin of life occurred throughout the long lifetime of Alexander Ivanovich Oparin, who himself published many books on the subject. For most of his career, Oparin was director of the Bakh Institute of Biochemistry housed in a large, multistory building on Leninsky Prospekt, in Moscow. The institute was known for its studies of fermentation and for the book-writing and conference-organizing activities of its director. The conferences nearly always provided a comprehensive treatment of some aspect of the origin of life.

Oparin drew some significant conclusions from his own experiments on coacervate droplets, such as the importance of organization in living units and the manner in which organization might have been achieved originally. As models for the first cells on Earth, coacervate droplets were, however, ill chosen; they obscured what they purported to explain.

Oparin recognized the instability of coacervate droplets quite fully and saw the need for bridging the gap between such slack structures and the robust cells of all organisms that we know. To overcome this conceptual defect, he invoked natural selection. For each successive descendant generation of spontaneous coacervate droplets on Earth, Oparin suggested, the most stable would be selected by the environment. These could then carry the properties to greater stability in the next generation.

This kind of thinking erroneously placed the onset of natural selection early in a series of precellular stages. When the thermal protein forming laboratory protocells came from the evolutionary direction, the resultant cells were found to be immediately stable, as well as far more catalytically dynamic than the gelatin used in coacervate droplets. Oparin's resorting to natural selection in order to introduce adequate stability into an evolving cell was shown by experiments on proteinoid microspheres to be a superfluous rationalization.

EIGEN

Manfred Eigen, a German Nobel laureate for physiochemical advances, has championed what he calls hypercycles.[11] Here we will simply quote from Robert Shapiro's *Origins*:

> A complicated and cooperative series of interactions, of checks and balances, developed between various nucleic acids and proteins. They have been named hypercycles and subjected to extensive mathematical analysis.

> . . . The hypercycles gained in complexity, and in their control of the environment, until a limit was reached. . . . Eigen and three co-authors closed a recent *Scientific American* article with the words: "The principles guiding the evolution of such an organization have been formulated and experimentally verified. Now what remains to be discovered is just what the favorable molecular structures were." . . . A Harvard biochemist, L. T. Troland (he was cited by Muller as a forerunner of the latter's thinking), wrote in 1914: "Consequently we are forced to say that the production of the original life enzyme was a chance event."[12]

Shapiro thus lines up Eigen, H. J. Muller, and Troland as proponents of the view that life began by chance, that is, from random beginnings. Agreeing with this point of view, Eigen has stated that biologically active proteinoids cannot be produced in the same way repeatedly. This would be correct if they were the products of a random synthesis from random mixtures of amino acids. Experiments repeated since Eigen made this statement show that his supposition is wrong—that polymerization is indeed reproducible. Eigen also predicates many ideas on the selforganization of RNA. However, RNA does not organize into cells, or into organelles, all by itself.*

CRICK

The contribution of Francis Crick to the theory of the origin of life is summarized in his book *Life Itself: Its Origin and Nature.* As we noted earlier, virtually all experts who claim to explain the origin of life propose a more or less detailed mechanism of the events involved—on the basis of experiments or conjecture, or both. Crick seems to *assume* that the mechanism required a first catalyst for the replication of RNA (earlier he proposed DNA). Since he indicates he does not see how the early Earth would have provided an appropriate scenario, he proposes that primitive life was brought to this planet by directed spaceship. This is "directed panspermia." That Crick explains the mechanism of the origin of life neither here nor there is a frequent criticism of his book.

*In a paper published in 1986, Eigen has changed his emphasis to a "value landscape that is not-random."[13]

Various Places, Various Ideas

ORGEL

The work of organic chemist Leslie Orgel is occasionally cited as an instance of experimental studies on the origin of life. The experiments he has reported are elegant examples of organic chemistry; it is probable that nonexperimental theorists like to refer to them for that reason.[14]

Orgel himself is more sober than his interpreters in assessing the relevance of his studies. At a meeting in Mainz in 1983, he reported his latest experiments, in which he produced polynucleotides on a template. Polynucleotide templates (guiding structures) are, however, synthesized by proteins (enzymes). The president of the meeting, Klaus Dose, put approximately the following question to Orgel: "You are careful not to claim too much. But others do make special claims for your experiments. So, I ask you, Where on the primitive Earth did the template come from?" To this where-on-the-Earth question, Orgel replied, "I have no idea." In a scientific meeting in Rome in 1985, he forthrightly characterized the relation of his experiments to the origin of life as "dubious."

More recently, Orgel proposed a transition from a mineral system to a polymer system through the interaction of organic phosphate with mineral phosphate. He candidly described the problems that needed to be solved and solutions that he proposed to work on as "flights of fancy," which at the time, however, seemed open to experimental study. Openness to experimental test can be claimed for a number of investigations that have not been done. Only a small fraction of the conceivable studies on the emergence of life have been tested experimentally. (This needs to be kept in mind; our subject matter is far from closed.)

On the positive side, it can be said that Orgel's experiments as well as those of Oparin and others help reveal relevant principles.

WALD

The theory of George Wald was first presented in 1954, a date that in many ways opened the present era of scientific discussion of the origin of life.

A number of Wald's ideas have been largely forgotten, partly to be rediscovered by others as if Wald had never expressed them. One author has said that Wald's classic 1954 article in *Scientific American* failed to mention the importance of the selforganization of matter in the theory of life's origin. In fact, the paper proposed this bright new idea and thereby changed the thinking in the field. One may instead ask, What major concepts have been advanced by other *theorists* that were not earlier put forth by Wald? How much can be forgotten or overlooked in the busy area of popular science in thirty years!

Wald's theories about the origin of life came naturally out of his rigorous experimental work on the chemistry of vision. His interest paralleled that of another well-known experimentalist, Melvin Calvin, whose main studies were on photosynthesis, the chemistry of green plants. These areas lead investigators to wonder how vision and photosynthesis got started. Both Calvin and Wald received Nobel Prizes for their respective studies. Wald is a biochemically knowledgeable physiologist and Calvin a biochemically knowledgeable organic chemist. Each has developed his own style of thinking and experiment, for objectives that were not in the mainstream of biochemistry.

Our area deals before everything else with the problem of how chemical occurrences modulated to biological ones. The subject matter components of the problem come naturally to biochemists, although until 1950–60 the vast majority of them seemed to shun it, for probably a variety of reasons. The first major biochemical textbook to devote a chapter to the origin of life was Lehninger's in 1970. Lehninger stated that in the preceding two decades the problem had moved out of the armchair into the laboratory and therefore deserved space in his textbook. Oddly enough, the majority of those who write books in the area and of those who do experiments in it, are not biochemists.

Wald grasped the seminal significance of molecules in getting life started. As he said, "given the right molecules, one does not have to do everything for them, they do a great deal for themselves."[15]

Wald saw also the importance of their ability to organize themselves. Here he took off from experiments. In the early 1950s, an illustrious cross-town colleague of Wald's, Francis O. Schmitt of MIT, reported experiments in which he allowed protein (collagen) molecules to precipitate slowly from solution. Schmitt's collagen formed beautiful microfibrils with regularly spaced cross-striations. This was a new observation in part because the electron microscope, used in photographing the fibrils, was itself new at the time. Schmitt's advance helped put the idea of the selforganization, or the self-assembly, of molecules on a firm basis. It opened up a new era in biochemistry and cell biology, with aid from Wald's placement of this phenomenon in an origin-of-life context. It is significant that twenty years later Lehninger's concluding chapter on the origin of life was preceded by one on morphogenesis.

Wald's overview and insight enabled him to see here a prime explanation for the origin of cells—the selforganization of protein precursors. To Wald, it was clear that cells are composed primarily of protein molecules.

His paper reported a number of other concepts that have become clouded by historical fog. A key idea was influenced by Oparin's view that natural selection occurred in precells rather than in protocells. This interpretation was the best available to Wald for his theory. It was another several years before experiments showed that thermal protein protocells already had the requisite stability at the outset (chapter 5); the invocation of natural selection in a prebiotic stage was accordingly superfluous.

The ideas that compose any one expert's concept of the origin of life are not a theory but a cluster of interrelated theories—what the physicist-philosopher Thomas Kuhn has called a paradigm.

Two main components of Wald's paradigm are (1) the primacy of proteins and (2) the idea of selforganization in the origin of cellular life. A third aspect had to do with whether the molecular matrix was random or whether it was determined by prior forces. We will discuss the answer to that question in chapter 8.

SHAPIRO

Robert Shapiro's 1986 book *Origins: A Skeptic's Guide to the Creation of Life on Earth* is notable for its unwillingness to accept ideas simply because they are widely held and venerable, and for its emphasis on experiments. A newcomer wishing to acquaint himself with the scope and vicissitudes of the concepts in this field can probably not do better than to read Wald's 1954 paper, Oparin's 1957 book, and Shapiro's 1986 book. The last is also recommended for critical surveys of ideas not discussed here, such as those of Hermann J. Muller, Sir Fred Hoyle, and Carl Sagan.

Shapiro, a biochemically oriented organic chemist, has become disillusioned with the view that nucleic acids came first, having himself experimented with and thought about nucleic acids. He goes into his reasons for inferring that proteins rather than nucleic acids had priority. In his criticism of nucleic acids first, Shapiro states, "Even the building blocks of nucleic acids, the nucleotides, are intricate molecules containing over thirty atoms each and requiring the precise connection of three subunits, with the release of two molecules of water. It is not surprising that prebiotic syntheses of nucleotides have run into intractable problems. These substances were developed well after life began."[16]

Following further arguments, Shapiro states, "Thus, the well-known Central Dogma of molecular biology, 'DNA makes RNA makes protein,' was exactly reversed in the development of life: in the beginning there was protein. Protein begat RNA, and then both begat DNA."[17]

The last sentence in his book reads, "We may be closer to the answer than we think."[18] In mid-1986, a reviewer of the book in the *New Yorker* stated that the nucleic acid–first view was no longer tenable; the answer had to come from some other starting point, perhaps proteins.[19] The tide of general thinking was beginning to turn, back to the days when T. H. Morgan had said to a young graduate student, "Fox, all the important problems of biology are problems of protein."[20]

Various Places, Various Ideas

Newer Data on the Self-Ordering of Amino Acids

After the appearance of Shapiro's book and well into the writing of this one, independent supporting evidence for the self-sequencing (self-ordering) of amino acids had not been published in the scientific literature. Such evidence was however being accumulated in the Max Planck Institute for Biochemistry, near Munich (chapter 8).[21] It allows us to infer that over a period of more than three billion years evolution, including protein synthesis at the center of the development, has resulted from the nonrandom process of the self-ordering of amino acids. For approximately thirty years, the proteinoid model was attractive primarily because it was the only one up to that time to answer numerous questions of emergence. It was also attractive because it yielded pictures of laboratory protocells that were included in numerous textbooks.

The independent supporting evidence for the nonrandom synthesis was seen to extend into other areas, and thus to provide a new perspective on Darwinian evolution. The findings in Munich further endorsed the inference that the variation step in Darwinian evolution was a nonrandom one. Moreover, this independent evidence came from the subject of biochemical evolution, manifesting itself from the inside of Darwinian evolution. Selection could now be viewed at the level of molecules (molecular selection) and at that of organisms (natural selection), as molecular selection offered new opportunities for the study of the variation step, and variation could yield to further investigation at a level of chemical and biochemical evolution.[22]

CHAPTER 8

The New Evolutionary Paradigm: Molecular Selection, Natural Selection, and Determinism

Darwin's theory of the origin of species by means of natural selection first appeared in print in 1859. Both its positive appeal for some readers and its repulsive shock for others were due to the fact that it removed the creation of species from the realm of the miraculous.

These two general reactions persist to this day. Indeed, most citizens who respond positively to the idea of evolution do so because they have long been bothered by the idea that the Book of Genesis is mythological, while those who respond negatively to the "very idea" of the theory do so because they find it to be utterly antibiblical.

In the realm of science, the theory has had a variable fate; most recently, it has been affected by an infusion of origin-of-life thinking. The general theory of evolution, however, holds the concepts of this book together as much as the spine and the glue hold the covers and the pages together.

Whether considered narrowly or broadly, the theory has itself evolved. For Darwin, evolution was a two-stage process, even though the title of his magnum opus, *On the Origin of Species by Means of Natural Selection,* seemed to indicate that the total process depended solely on selection. In his introduction, Darwin stated, "I am convinced that Natural Selection has been the most important, but not the exclusive, means of modification." In his conclusion, he forecast, "A grand and almost untrodden field of inquiry will be opened, on the causes and laws of variation."[1]

For Darwin, it is evident on closer examination, evolution was mainly but not merely natural selection. The essential processes were variation, competition, and selection, not to mention reproduction and inheritance. It is not always clear which aspects in this group are truly processes and what the other aspects are. Darwin's discussion of variation was extensive. Perhaps he gave the earlier step of variation second place in the evolutionary mechanism because he was in his time unable to analyze variation in the thorough manner that characterized all of his scholarship. Despite his greater emphasis on selection, Darwin recognized it as a passive process, not as the active one many of his modern-day disciples have made it out to be.

Darwin saw evolution by selection as a gradual process. The disciples who came later, the neo-Darwinists, were not of a single mind, but they relied on gradualness even more than Darwin did. They tended increasingly to interpret the process of natural selection itself as a driving force.

In the century after Darwin published *On the Origin of Species* doubters came and mostly went. Beginning in 1965, the dissension increased anew, slowly at first. It focused on the generally accepted source of variation. The common belief was that such variation was random, that it arose by chance. Often, but not always, this variation was, in fact, of a random, playing-card variety.

This view grew its roots in population biology and was strengthened in the late twentieth century by physical scientists who began to look at the theory of evolution. These included Jacques Monod, Manfred Eigen, and Francis Crick—all Nobel laureates and all quite influential shapers of general thinking. In his book *Chance and Necessity,* Monod said, in referring to protein, "In its basic make-up it discloses nothing other than the pure randomness of its origin."[2] In the same year, 1971, Eigen said, "In the beginning . . . there must have been *molecular chaos* . . . the origin of life must have *started* from random events. . . ."[3] In his book *Life Itself* (1981), Crick had this to say:

> My own prejudice is that nucleic acid (probably RNA) came first, closely followed by a simple form of protein synthesis . . . unless there were some rather specific catalyst present. This conceivably could be a mineral or even some peptide produced by the random aggregation of amino acids.[4]
>
> This does not mean that accidental polymerization might not have produced proteinoid molecules which might perhaps have assisted in the buildup before true replication finally occurred.[5]

Here Crick thus acknowledged the proteinoid alternative in the course of reiterating his belief that nucleic acids arose before proteins.

Crick looks at the process of evolution from the outside and tends to ignore an overview of chemistry on the inside. Crick is here equating the process of aggregation, the lining up of monomers, to polymerization. Polymerization, however, is a second step, following the attraction of the monomers to the polymer. It requires their actual bonding. Bonding needs an input of energy, and it nearly always uses catalysts. Crick furthermore takes for granted that both steps, between which he does not distinguish, are random. He simply assumes first that the aggregation of monomers is random and then that the subsequent step of bonding during polymerization is also random, as indicated by his use of the synonym *accidental*.

For the most part, these authors, whose statements are being cited as illustrations of widespread thinking, were reporting their thoughts on the sequence of the beginnings of organic evolution leading to later steps. When we turn to someone who focused almost exclusively on the earlier steps, we find early reinforcement for the randomistic, order-out-of-chaos view. A chief proponent of this approach was Oparin. His statements antedate those cited above and may indeed have had considerable tacit influence on them.

In 1957, Oparin said we cannot expect to do more for the understanding of stages on "primeval Earth" than to

> . . . explain the formation of organic polymers in the shape of polypeptides and polynucleotides, assemblages having, as yet, no orderly arrangement of amino acid and nucleotide residues adapted to the performance of particular functions. . . . It is only by the prolonged evolution of these systems . . . that there developed metabolism, proteins, nucleic acids, and other substances . . . which characterize the contemporary living organisms.[6]

Ideally, there should be no authorities in science. But there are individuals who are of unusual importance because their statements are much listened to; they are in fact authorities. The four authorities just cited are only representative. Their views are close to the general consensus. They differ from the average views of biological evolution because their authors are themselves physical scientists, whose concept of randomness has been reinforced in their own area by interpretations of indeterminism, stemming from famous physicists like Heisenberg and Bohr.

These various ideas are paradigmatic. Randomness and the nucleic acids–first concept are not immediately apparent as being mutually dependent, but close examination indicates that they are. That stems from the general assumption—expressed by Monod, for example—that the first proteins were random, the result of chance. RNA thus had to be there at the outset to select proteins to give the limited varieties we see in nature. The reasoning is that ordered proteins required RNA because they would otherwise have been random, virtually unlimited.

Views of Randomness in Evolution

Of the scientists quoted above, Monod could claim to be a biologist's biologist; Eigen and Crick are primarily physical scientists. How do conventional biologists look at randomness as a basis for evolution?

A famous and representative evolutionist, C. H. Waddington, stated, "Evolution is brought about by the natural selection of random variations which occur in the genetic material."[7] Salvador Luria, a biologist who won a Nobel Prize for the studies he and Max Delbrück did on mutation in bacteriophage, also feels that the "process of genetic mutation is strictly random."[8] Luria explains that order is rescued from randomness by the process of selection.

Problems like the origin of life produce special interactions across the borders of two sciences. Biologists tend to favor definitions of randomness that are less rigorous, more yielding, than those of the physicists. Biologists lean blindly on physicists for some definitions, while physicists are not likely to devote extensive amounts of time to reviews of descriptive detail, as they correctly believe biologists do.

When different scientists tackle a problem that concerns both extensive detail and searched-for principles, the problem takes on a life of its own. Ultimately, the history of the theory of evolution will need to reveal how physicists, chemists, geologists, biologists, and others each think and how they all think. The multiple requirements of these multiple lines of thought, fortunately, introduce rigor.

The Beginning of Dissent

One of the most dramatic challenges to the modern Darwinian theory was initiated by Victor Weisskopf, head of the Department of Physics at MIT. In 1966, Weisskopf gathered some of the theory's doubters and supporters, and the group presented a balanced treatment of the subject at the Wistar Institute of Philadelphia under the title "Mathematical Challenges to the Neo-Darwinian Interpretation of Evolution." Since the base of neo-Darwinism was mathematical probabilism, it was appropriate that the challengers were physical scientists familiar with the mathematical base.

The leader of the Wistar Institute opposition to neo-Darwinism was Murray Eden of MIT. He expressed his dissent with a penetrating analysis of the concept of "random":

> Any principal criticism of current thoughts on evolutionary theory is directed to the strong use of the notion of "randomness" in selection. The process of speciation by a mechanism of random variation of properties in offspring is usually too imprecisely defined to be tested. When it is precisely defined, it is highly implausible.[9]

Although the Wistar Institute meeting challenged the established thinking in Darwinian dialectics, the full significance of various comments and papers became manifest much later. Meanwhile, Eden and others supported their criticisms with mathematical analyses.

Five years later, in a book devoted to molecular evolution, Szent-Györgyi presented a nonmathematical criticism of randomness in evolution. He took into account new views on how biologically significant compounds first came into existence. He is worth quoting both on the origins of life and on Darwinian evolution:

> Not long ago evolution was shrouded in an impenetrable mystery. We thought that more complex molecules could be built only by living systems. But the living system itself is built of such molecules, and how could a system build its own building stones? How can a hen lay the egg out of which it has to be hatched? This paradox is now partially understood. We know that Nature can build complex organic molecules without the intervention of life. . . .[10]

For Szent-Györgyi, Nature was the geological realm that contains organic compounds and is able to use energy for further organic reactions in making complex molecules.

From this overview, he proceeded to the emergent problem of evolving life:

> Another paradox still awaits solution. Mutation is a random process. How could a random process lead to such highly ordered structures as a multicellular living organism? The usual answer to this question is that there was plenty of time to try everything. I could never accept this answer. Random shuffling of bricks will never build a castle or a Greek temple, however long the available period. A random process can build meaningful structures only if there is some kind of selection between

> meaningful and non-sense mutations. A certain selection was already supposed by Darwin, who made the struggle for life responsible for selection. The selection, however, in my opinion has to take place much earlier, not only with the final product.[11]

While Eden objected to neo-Darwinism on the basis of mathematics, Szent-Györgyi's doubts stemmed from techniques of biochemical analysis.

Almost two decades later, the challenge was renewed by younger scientists. The editors of Academic Press in London invited Mae-Wan Ho and Peter Saunders to assemble a group of physicists, chemists, and biologists to explain the growing unease with neo-Darwinism. That such a book was published in England, the land of Darwin, seemed quite fitting.

The editors of Academic Press did not go to the scientific establishment of Cambridge University for the major part of this task, although the book does contain a supportive foreword by Joseph Needham of Cambridge. Mae-Wan Ho and Peter Saunders, her husband, are British subjects, but each came from elsewhere. Ho holds a Ph.D. in biochemistry from the University of Hong Kong, while Saunders is from Canada. His skill in administering departments of mathematics is such that he was at one time assigned to head simultaneously three of them in the University of London complex.

In their critique, Ho and Saunders summarize, "The neo-Darwinian concept of random variation carries with it the major fallacy that everything conceivable is possible."[12] This book set off an intense discussion, which is continuing. Elisabeth Vrba, a paleobiologist from Pretoria, reviewed the fossil record and made a plea for a more complete theory. In a summary statement in a popular science magazine one year later, Vrba was quoted as having said, "Most biologists are still transfixed on natural selection acting on random mutations as the mechanism of evolutionary change. But natural selection may well shrink in importance and we could find that mutations are anything but random."[13]

It is helpful to compare the new evolutionary paradigm with the old one. In a sense, there never was an old paradigm: the one that existed was dominated by theoretical exercises from the outside. Evolution, however, began on the inside and continued on the inside, and by selection acting in the environment, on the outside of organisms. Since it really began with the Big Bang, it began on the outside—but the outside of what? Since we arbitrarily limit ourselves to the origins and evolution of organisms, we can say it began on the inside of organisms. What could be thought about evolution was based on analysis of what is here now and on a kind of extrapolation backward, in which it was not possible to say what pattern, if any, was set by the origins. Now we know that the initial pattern was crucial.

The only way for us to retrace origins is to position ourselves before origins and then retrace forward (see chapter 1). This required experimenting heuristically, as nature herself did. Important principles have emerged from this approach. Until evolution was studied that way, an evolutionary paradigm, strictly speaking, did not exist.

One other critic may be quoted—Gordon Rattray Taylor, author of *The Great Evolution Mystery,* published in 1983, the year of his death. Taylor was the chief science adviser to BBC television and author of sixteen books, including *The Biological Time Bomb.* He did not break with the establishment on the issue of randomness, and he correspondingly assumed that biotic order emerged from prebiotic chaos, but his analysis of the theory of evolution in his time is masterly, and his conclusion is an epigram: "Darwinism is not so much a theory as a subsection of some theory as yet unformulated."[14]

Further formulation came from physical scientists, especially those few chemists interested in evolution. Ernst Mayr did not see it that way. In his grand survey, he stated that progress was not made until the field was emancipated from the physical scientists. He saw that emancipation as having come at a time when a number of evolutionists themselves were inviting physical scientists into their domain. On this, I can speak from personal experience obtained from the inside.

The National Science Foundation operated a grants panel in systematic biology (taxonomy), the branch of biology that systematically classifies plants and animals. In the late 1950s, the experts on this panel became aware that they needed to include an evolution-minded chemist to help them judge proposals based on chemical techniques. By this time, chemical techniques had come to the fore in helping to classify organisms. The panelists wanted a chemist sympathetic to systematic and evolutionary exercises. My invitation to serve was no doubt related to my having published a few research papers in which, as a chemist, I had tried to explain aspects of evolutionary processes. My service on the panel was to help form judgments on the feasibility of proposed chemical analysis used in taxonomic studies. It turned out to be an excellent educational experience for me, since I learned how eminent panelists like Al Romer, Libby Hyman, Rogers McVaugh, and David Keck thought about systematic biology and the mechanism of evolution. It also suggested how eminent evolutionists who submitted research proposals did much of their thinking.

During one panel meeting in Washington in 1959, I asked the eight members present how they would vote if Charles Darwin came in with a proposal *before* his theory was known. The frank response from most was that they would reject his proposal as too visionary. Two or three others said they thought they would probably let him have trial funds for a year or two to see what he might come up with. My question was somewhat unfair in that a thorough judgment would have needed a completed proposal from Darwin. The responses illustrate, however, that although the system supports sound research, it is unsure what to do with truly innovative studies. At the time I put my question, I did not ask what the panel would do with a broad proposal from a chemist interested in the topic. It was another six years before I even participated in the Wistar symposium on evolution and fifteen years before I suggested a theory of the beginning of evolution from the chemical inside. But the impetus to think about evolution in chemical terms came even more from this exposure to the panel's activities than from earlier conversations with Morgan.

Molecular Selection and Natural Selection

As long as the understanding of variation was predominantly descriptive from the outside, not much could be said in specific terms. Since the early 1960s, the major emphasis in the Darwinian paradigm has been on the action of natural selection on random, non-goal-directed variants. For neo-Darwinists, the initial range of variants was random, and chance played a large role in continued variation. They, somewhat unlike Darwin, tend to see natural selection as a probabilistic and a primary directive agent.[15] They visualize the neo-Darwinian theory as the grand synthesis. The modern dissenters, quite like Darwin, believe that something important has been missing from the theory, especially an understanding of the variation step.

Some of the participants in the Wistar conference of 1966, many contributors to the Ho-Saunders book, and other critics, such as Szent-Györgyi, believe that what is missing is the limiting conditions of nonrandom processes. This view grew from experiments retracing the formation of prebiotic protein from its constituent amino acids. The process is a highly nonrandom, highly deterministic one, sometimes referred to as molecular selection. It is the active process within evolution. Natural selection serves to winnow out the less successful molecular or biological variants, but only after a great deal of internal molecular selection has led the way.

In his magnus opus *The Growth of Biological Thought,* Ernst Mayr traces the development of the evolutionary synthesis of 1980. He attributes it to "a heightened confidence in the powers of natural selection" and to a handful of conceptual bridge builders.[16] This development might have made Darwin happy, but it failed to build a bridge between the physical aspects of prebiology and the exterior aspects of organismal science. Mayr had earlier described this gap as unimportant. He and other neo-Darwinists invoked genes for internal aspects of variation, but it is certain that T. H. Morgan, a biologist who emphasized a deterministic universe, would have found that aspect of neo-Darwinism unacceptable. Molecular selection, though not called that, nor analyzed, was a cornerstone of Morgan's thinking.[17]

A CONNECTION BETWEEN THE ORIGIN OF LIFE AND ITS EVOLUTION?

A special hurdle for the critics of the Evolutionary Synthesis has, despite Mayr's comment on bridge building, been the argument that the origin of life and the theory of evolutionary biology are separate subjects.

If it be correct that the processes and phenomena of life's origin and those of its evolution are independent of each other, the principles of Darwinian evolution are self-contained and the comments of Szent-Györgyi, Taylor, Ho and Saunders, this chapter, and so on are all irrelevant.

Prior to the development of a tangible basis for a decision on this issue, there seems to have been no motivation other than faith to mesh origin of life with evolution of life. One has faith either in an orderly universe or in a probabilistic, chancy universe.

That the two subjects are independent is a view still widely held. Their separation even reached the status of a 1982 legal opinion, in a trial on the teaching of evolution in Arkansas. While the decision not to allow the teaching of the Book of Genesis in science courses in Arkansas is a proper one, the legal opinion includes the statement "Although the subject of origins of life is within the province of biology, the scientific community does not consider origins of life a part of evolutionary theory."[18]

This legalistic statement was undoubtedly based on advice from several scientists in attendance at the Little Rock trial. Their assessment of the thinking of the "scientific community" was largely correct, but it did not allow for a minority opinion. The scientists were selected by a legal staff of the American Civil Liberties Union working with lawyers pro bono from a private law firm.

The point is that a large part of the biological "scientific community" does think as Judge Overton's advisers suggested. These latter are pretty much well-established scientists, and the newer minority point of view was shut out from a legal opinion, in a subjective and cavalier way. The newer thinkers, such as the authors in the Ho and Saunders book, and the editors, Ho and Saunders, in their *Beyond Neo-Darwinism: An Introduction to the New Evolutionary Paradigm,* and Stanley Salthe, a professor of biology at Brooklyn College, among others, had seen that Darwin's theory needed to be integrated with the protobiological matrix that was illuminated only long after Darwin's time.

The New Evolutionary Paradigm

Although not foreseen, the results of experimental protobiological retracement indicate that origins *do* interdigitate with the early evolution of organisms.

This evidence consists, first, of such interpretations as that the path from prelife to life, as we know it, passes through protein. As Szent-Györgyi emphasized, protein can be made in the geological realm without organisms. When it is chemically basic, it can to a degree make a more modern kind of protein in much the same way modern organisms do now. The properties of the later-generation proteins need to be studied, but we have reason to believe that they are also archetypal. In the making of these more modern proteins, the energy comes from ATP instead of heat.

Second, the aggregation products of those thermal proteins are indistinguishable from the fossils of unicellular forms in the most ancient fossiliferous strata on Earth. The fossil evidence at the cellular level is thus as supportive as we can expect it to be.

Third, the development of this model has been synonymous with that of all cell construction science. This largely, but not entirely, new discipline consists of (1) the noncellular synthesis of protein, followed by (2) the selforganization of the proteins. Construction of cells lies outside the mainstream of modern science, but it is very much in the flow of evolution itself. Evolution is forward and, except for short stretches, irreversible.

Fourth, principles found to be central to protobiology apply both to prebiology and to biology.

Independent evidence for the total picture of evolution from inanimate matter through the first cell and beyond, into the mechanism of an updated Darwinian evolution, appeared while this book was being written. The new data stemmed from a laboratory model in which biological information, as already described, is seen by experiment to originate in abundance from an ancient, nonrandom reaction of several or more types of amino acids. The primary step, according to the picture, was the amino acids' ordering themselves during that reaction. This process had a geological simplicity. Nothing anywhere near as complicated as the modern genetic coding mechanism was necessary to put into the evolutionary stream the biological information that the present stage of evolution tells us must have arisen at some time.

The guiding principle, then, was the self-ordering of amino acids. The model that resulted in ordered protein, enzymes, cells, reproduction, and numerous other properties of modern living things was nevertheless still a model. While the textbooks reported the results of the model abundantly for twenty-five years, none could go so far as to assert that an original thermal protein was the first ancestor of all the later proteins. A related difficulty was that the leap from then to now spanned more than three billion years. A new kind of evidence was desirable, if it existed.

The independent source of evidence was the laboratory of Gerhard Braunitzer, a famous German protein chemist at the Max Planck Institute for Biochemistry, near Munich. Two of his younger colleagues, Orlin C. Ivanov and Berthold Förtsch, applied a newly advanced method of computerized analysis to all of the proteins available for study in 1984—2898 of them, a record number to be studied. They found abundant evidence that these modern proteins contained evolutionary relics of an original self-ordering of amino acids. They referred to these relics as "universal." This meant that alongside the genetic coding mechanism, which was still controlling, the modern cell has also a continuation of self-ordering of amino acids. Moreover, the operation of the ancient component was more manifest in those proteins that bore the marks of being the first ones to join the parade of evolution.

In the final paragraph of their discussion, Ivanov and Förtsch reviewed the reasons for their not accepting the idea that the governance of the amino acid sequence in modern proteins is the function of DNA/RNA *alone;* they also summarized the evidence for the self-ordering of amino acids. In the last sentence, the authors state, "This is in accordance with the 'proteins first' conception because at that stage protein structure must have been determined only by preference in bonding [self-ordering] mechanism."[19]

Because of this evidence alone, the mechanism for the conversion of inanimate proteinoid matter into living things, from a unique laboratory model, has come to be considered a general process operating through more than three billion years of evolution. This broader view also increases the likelihood that the principles of stepwiseness, self-ordering, and selforganization extend into the biological realm.

Analysis indicated that resistance came largely from neo-Darwinism in biology. Later, game theory in physics was seen to reinforce neo-Darwinism.

Intensive examination of the structure of neo-Darwinism indicated that this popular paradigm favored randomness and indeterminacy in some form. The proteinoid paradigm, on the other hand, was (and is) fundamentally an expression of nonrandomness.

The recognition by far-seeing bioevolutionists of the pervasiveness of nonrandomness made its way into the literature. Ho's call for a chapter and a book to correct this view was a turning point.

Together at Last (A Recapitulation)

As long as the theory of emergence of life was undisciplined by experiment, it remained essentially unintegrated with the theory of Darwinian evolution. Theories were developed separately in the two

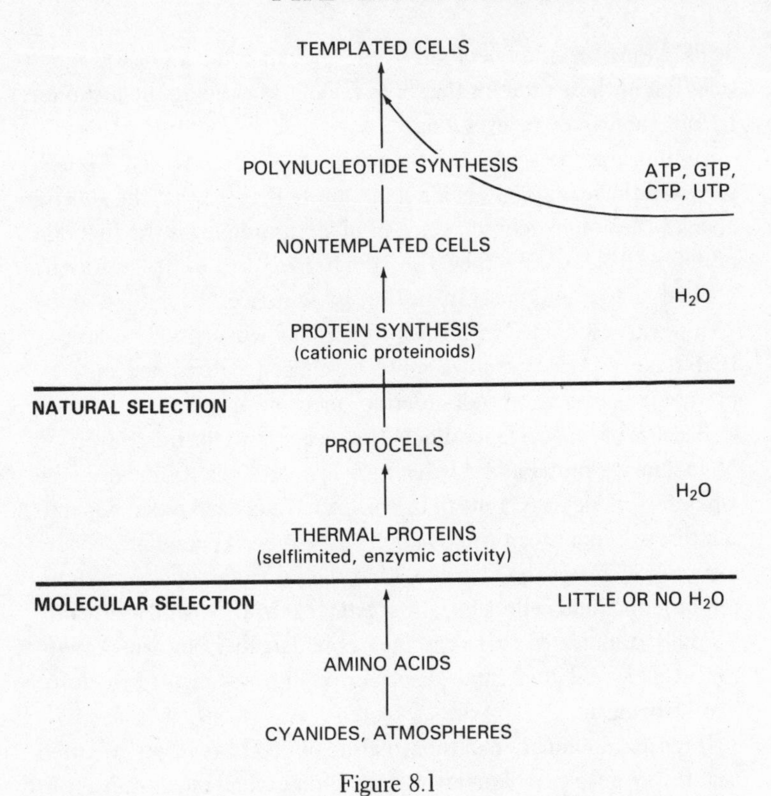

Figure 8.1

Flowsheet of the proteinoid (thermal protein) emergence of life leading into its evolution. The self-ordering of amino acids (molecular selection) has made it possible.

areas. We can understand this best when we recall that Darwinian theory was necessarily built on biological description from outside the organism, whereas theories of the origin of life were necessarily built on the testing of chemical processes from inside the first organism-to-be. As we have seen, Oparin attempted to bridge the two by invoking Darwinian selection for the development of stability in the first cells. However, Oparin's coacervate droplets failed to call attention to the orderly, deterministic connection from prebiotic events; this was later suggested by informed thermal protein and its selforganization into microspheres. In this newer context nonrandomnesss was and is at the center of the connection between prebiotic and biotic.

Albert Lehninger, later than Oparin, supposed that a theory based on proteins first could yield integration. In the 1975 edition of his very popular textbook *Biochemistry,* Lehninger portrayed the sequence of chemical evolution entering into Darwinian (biological) evolution. The flowsheet in figure 8.1 is a partial updating of Lehninger's schema for evolution, in which protobiochemical events and stages of evolution are integrated. This flowsheet is also a schematic summary of processes discussed in this book.

Lehninger perceived protein as a connecting link between prebiotic and biotic substances, as does this volume. The connecting principles were two kinds of limitation, or selection. In Lehninger's chemical evolution, what we call molecular selection was a central informational process, while in the ensuing Darwinian evolution the selection was natural selection.

In their dissatisfaction with neo-Darwinism, which favors natural selection from random beginnings and random variation, Ho and Saunders explained the new view further, based on nonrandomness. The consequences have been several: (1) further corroboration of the nonrandom model of emergence; (2) at least partial filling of the gap in Darwin's theory as requested by numerous theorists, including Darwin himself; and (3) the crystallization of a larger theory that extends beyond the old boundaries.

Randomness Versus Determinism

DETERMINISM

The roots of the new evolutionary paradigm are at the fundamental level of chemical reactions. Although chemists characteristically do not think in philosophical terms, the analysis to this point suggests that the essential kind of evolutionary control has profound philosophical implications.

In everyday terms, *determinism* signifies fate. Large parts of the human population believe in fate. We encounter that belief in common sayings like "What was meant to be was meant to be." Divinely controlled fate is at the heart of the Islamic religion. Most other religions also rely on divine determinism.

The scientific view of determinism acting through the evolutionary hierarchy features two themes: the popular and negative one of randomness, the indeterminate or probabilistic emphasis; and the nonrandom one—a theme much more difficult to score mathematically than a probabilistic one.

Yet proteins come not out of playing cards but out of nonrandomly interacting molecules. The experimental results suggest that we need to look for our fate not to playing cards and game theory but to molecules. It is molecules that initiate, "design," and "direct" evolution.

The modern paradigm tells us that instead of "It's in the cards" our guideline should be "It's in the molecules." From nonrandom arrays of protein molecules, we have begun to trace all other phenomena of interest to us as humans.

We need to remember the primacy of molecular evolution within a larger evolution. The popular perception is that chemical reactions are carried out by chemists. That may be another manifestation of anthropocentric ego. Chemists do not cause chemical reactions; the molecules engender them. In evolution, chemical substances are the prime movers. Evolution finally attained a level at which the chemist emerged, to state that he performed the chemical reactions. At the beginning and at the most recent stage, all is traceable to the appropriate molecules.

Missing from essentially all of the theoretical paradigms on the origins of life is the answer to the question found in the primary and inner nature of evolution: Is it random, or is it nonrandom, that is, is it deterministic? The answer provided by experiments, already mentioned, will here again be reviewed in the light of the preceding parts of the book, especially the catalog of various propositions. Such lists are necessarily incomplete; Shapiro missed Calvin's contributions, Oparin wrongly belittled Herrera's findings, a review of Wald's 1954 conceptual contributions overlooked his major emphasis on selforganization, and so on. This book also cannot claim to be complete. However, the experiments it reviews emphasize what appears to be missing from other treatments: a recognition of the experimental demonstration of determinism in the evolutionary sequence, and its significance.

To appreciate how pervasive the randomistic thinking is, we may look at its history. Virtually all of the commentators on the origins of life have taken the idea of randomness so literally that they do not question it. But it has been questioned—above all by Albert Einstein, who spent his later years in the back room of physics, because he could not agree with certain assumptions of quantum mechanics. This was ironic, given Einstein's large initial role in the development of quantum mechanics.

On this issue, highlighted by Einstein's thinking, we can see that life is a child of the universe and that we are not mistaken in making what at first looks like a huge leap in our thinking. The "leap" is really the climbing of a ladder larger than that from molecules to societies and is one having innumerable steps, many of which have been identified. We can hardly leap from the bottom rung to the topmost rung of even a small ladder. A long ladder of innumerable rungs can be very formidable, whether it is a ladder of wood or a ladder of concepts. Even so, we can climb it in steps.

The concept of randomness has a long history. The tides of scientific thought that concern us in our view of life's origin are found in our understanding (a) of the universe in which life emerged and (b) of the biota that emerged during the evolutionary diversification of original life.

In the popular picture, life arose in a disordered universe ("order out of chaos") and evolution depended primarily on natural selection acting on undirected variants (neo-Darwinism). The physicist's concept and the biologist's concept reinforce each other through their common emphasis on a dominating indeterminism. To understand this thinking, and what experiments indicate about it, we scan some of its history.

Historical Approach to the Ideas

In physics, the name of the game, or rather of the opponents in the game, are causality versus indeterminism. The belief in indeterminism begins with randomness; according to this view, effects arise out of statistical happenings. In biology, those occurrences that happen to contribute to survival in a specified environment are "selected." One question is whether the selection is natural or supernatural. In either case, an evolution that depends on selection from a statistical array appeals to a sense of logic.

Many histories of such questions begin with Isaac Newton and his *Principia.* Newton is often considered the scientist who put causality on a working basis through the physical laws that he enunciated. In the 1978 book *Isaac Newton's Papers & Letters on Natural Philosophy,* Thomas S. Kuhn stated, "Newton was guided throughout his scientific career by the conception of the universe as a gigantic machine."[20] Newton was the original mechanist. He left unanswered the question of whether the original cause was a natural or a supernatural one.* Much has been made of Newton's view that his science proved the existence of a supernatural deity; this view had a great

*A way of resolving this question for all except literal interpreters of bibles (there are more than one) is to say, much as Einstein did, that science is defining God.

Another way of regarding this question that has been considered is that of the poet William Herbert Carruth (1859–1924):

> "A fire-mist and a planet
> A crystal and a cell
> A jelly-fish and a saurian
> And caves where the cave-men dwell;
> Then a sense of law and beauty
> And a face turned from the clod—
> Some call it Evolution
> And others call it God."[21]

impact on thinking in the eighteenth, nineteenth, and early twentieth centuries. Indeed, the "scientific creationists" in the latter half of the twentieth century claim Newton for their own. In the book mentioned above, however, Perry Miller explained in exquisite detail how the efforts of a zealous interpreter, the clergyman Richard Bentley, were largely responsible for turning this interpretation of Newton to ecclesiastic profit, both financial and propagandistic. Miller saw Newton not as a theist or as an anti-theist but as a tool of the zealous and tendentious interpreter Bentley.

In any event, Newton was the famous early determinist, one who saw a cause for every effect. Even before him, Galileo had been an important experimenter and a causalist. So, in our own century, was Einstein. Modern science was founded by men like Newton and Galileo and brought more up to date by Einstein.

While nowadays the concepts of probabilism and randomness are enjoying ascendancy, supported by famous scientists like Max Planck, Niels Bohr, Paul Dirac, and Werner Heisenberg, determinism has had its day. These opposing points of view came to a head with Einstein.

Although Einstein's theory of general relativity has fared very well through the decades, he resolutely opposed until the end of his days the statistical interpretation that modern quantum theory stresses and which, not just coincidentally, supports neo-Darwinism. Critics on the side of molecular morphology sometimes say it is all a hypnotic fascination with numbers. Einstein's famous statement was "I cannot believe that God would choose to play dice with the world."[22]

That Einstein recognized his own intransigence is clear from a comment in a letter published in *The Correspondence between Albert Einstein and Max and Hedwig Born, 1916–1955.* The book contains notable forewords by two of Einstein's friends, Werner Heisenberg and Max Born, who were also his opponents in the area of quantum science. Heisenberg's words, in fact, are so moving that even the strongest partisan of Einstein's point of view must cheer for Heisenberg's sense of humanity. Heisenberg stated,

This correspondence should not only be rated an extremely valuable document in relation to the history of modern science; it also bears witness to a human attitude which, in a world full of political disaster, tries with the best of intentions to help wherever possible, and which considers love for one's fellowmen to be fundamentally of far greater importance than any political ideology.[23]

Einstein wrote to Max Born in 1944 as follows:

We have become Antipodean in our scientific expectations. You believe in the God who plays dice, and I in complete law and order in a world which objectively exists, and which I, in a wildly speculative way, am trying to capture. I firmly *believe,* but I hope that someone will discover a more realistic way, or rather a more tangible basis than it has been my lot to do. Even the great initial success of the quantum theory does not make me believe in the fundamental dice game, although I am well aware that our younger colleagues interpret this as a consequence of senility. No doubt the day will come when we will see whose instinctive attitude was the correct one.[24]

Einstein had held the same view for at least fifteen years. In 1929, he said in an interview,

Everything is determined, the beginning as well as the end, by forces over which we have no control. It is determined for the insect as well as for the star. Human beings, vegetables, or cosmic dust, we all dance to a mysterious tune, intoned in the distance by an invisible piper.[25]

This was a declaration of determinism about as total as one can expect it to be stated. The determinist argument brings up old questions of whether man has free will and, if so, how much. Is man a machine or, as Albert Szent-Györgyi put it, a rationalizing animal who likes to regard himself as a rational animal? Proponents of determinism point out that an increasing understanding of evolution causes man step by step to surrender his anthropocentric ego. This seems to conform with an observed trend of evolution itself—toward increasing independence of the environment. If the evolving individual is indeed seeking to be independent of the environment (because of some inner drive), his brain is likely to reflect the history of that intergenerational attempt.

At any rate, Einstein held to his belief in determinism, originally embraced before 1929, until the end of his days, in 1955. We have no evidence that he ever found the "tangible basis" that he sought for his beliefs.

The history in this area is marked by quarter-century turns in thought. Another quarter-century after Einstein's death, this question was examined afresh. A number of physicists have quietly doubted the largely indeterministic point of view. Of these, Koichiro Matsuno was unusual in looking actively for Einstein's "tangible basis."

Matsuno earned his Ph.D. at MIT in 1969; he accumulated many questions not answered in class. After that, he worked in Japan at the big Nippon Electric Company and later became a professor at the Technical University at Nagaoka. There he teaches physics to students and English to other faculty members, because of his familiarity with the language. Ordinarily, scientists think of physics as contributing to biology, but Matsuno thought in the opposite direction—biology to physics or, even more esoterically, protobiology to physics.

For Matsuno, the tangible basis was the evidence that the first protein precursors of life on Earth were not random, so that the first cells were not part of the "fundamental dice game" in Einstein's phrase. The "tangible basis" is the nonrandom condensation of mixed amino acids to yield primordial proteins, already formed and informed by inner and prior processes. When the young Japanese physicist explained this view in 1984 in the book *Beyond Neo-Darwinism*, the doyen of chemical embryology at Cambridge University, Joseph Needham, asked in the foreword, "Might he not have been influenced by the neo-Confucian ideas of East Asia?"[26] Matsuno's views, and Einstein's as well, are in tune with Eastern philosophy. His focus on the answer to Einstein's question has made Matsuno a master of the physical perspective on related areas, such as origin-of-life and bioevolutionary theory.

The proteinoid experiments emphasize nonrandomness. As Robert Shapiro stated in 1986, the neo-Darwinists and the San Diego creationists, and others, have been united in opposing the results of the proteinoid experiments. The creationists argue repeatedly that evolutionary theory is wrong because of its reliance on a random evolution, for which they then substitute divine direction. They are at least partly correct—on the issue of randomness in evolution. The remainder of the disagreement can be said to be over how one defines God. This argument is thus not properly between a mythological creation and a natural one but between two points of view within science. The indeterministic point of view was developed by physicists, despite its strong opposition to some of Newton's assumptions and despite a continuation of the conceptual conflict later, with such famous physicists as Bohr, Born, and Heisenberg on one side and Einstein and Erwin Schrödinger on the other. Randomness was appropriated by evolutionists and became increasingly dominant, until the dissenters became significant in number.

Until the realization that nonrandomness applied alike to the first steps in molecular evolution and to the events in biological evolution, most evolutionists separated origin of life from evolution of life. The more modern synthesis states that the informed life that originated became in steps of further evolution the continually changing organism. However, it retained throughout evolution its original protein makeup and synthesis mechanisms, its ability to synthesize its own template, its fundamental cellular shape, its ability to provide separation within and without by means of membranes, and the platform of an evolving metabolism.

Was Einstein Right?

Two of the most significant theories of the century have been those of relativity and quantum science. The verdict on relativity is in, as much as a scientific verdict can be. It is highly affirmative. Responding to a judgment on Einstein's relativity, another famous physicist, Heinz R. Pagels, stated, "Of course Einstein was right! . . . Over the last quarter century general relativity has been put to the test of experiment."[27]

On quantum science, where Einstein held what came to be a heretical opinion, the verdict is not in. The experiments, the "tangible basis," come from what to many physicists is a very unlikely source, model experiments on the emergence of protolife. The essential positions to be compared do not ease the problem: total determinism, "almost" determinism, "almost" indeterminism, and total indeterminism.

Einstein adhered to a total determinism. Even though he knew of and wrote on natural selection, there is no evidence that he thought of that process as modifying determinism. His friend the Nobel physicist Schrödinger did write on modified determinism, albeit without invoking natural selection.

The essential concept of modified determinism poses the question of whether there exists an element of statistical variation, chance, operating in a primarily deterministic evolutionary sequence. Darwin indicated in one of his notebooks that he believed that free will and chance are synonymous. Life and its evolution could theoretically be largely determined, and yet leave some room for the operation of human selection with a minimum of encumbrance built in by the evolutionary process.

For this kind of "almost" determinism, two main possibilities exist. In one, the scope of variation, and perhaps more than the scope, is also predetermined. In the other, thinking from physics and biology may be fused. For this, the importance of natural selection has to be believed in context. Natural selection has gained wide acceptance. It has, however, been accused of being tautologous, of claiming, for example, "That survives which is most fit to survive." Darwin's treatment of natural selection was couched in a larger framework of variation. Although he did not and could not, in his time, understand the "laws of variation," he explained that the process acted on those variants that increased survival or survivability. In order for natural selection to operate, there had to be variants from which the most survivable were selected and preserved. If there had been no variants, evolution as a process would have been paralyzed.

To suggest a total, or "hard," determinism, instead of a qualified, or "almost," determinism, seemed justified in Einstein's time and still seems so in the last part of the twentieth century, simply because the prevailing paradigm has long been far on the side of indeterminism. As we noted, Einstein endured powerful opposition.

So, according to the experiments life, collectively and individually, is close to an expression of determinism. The initial chemical reactions of amino acids were close to deterministic, the array of informed thermal proteins serving as the doorway between the inanimate and the animate was close to determined, and the generation of cells was highly determined.*

This internal directedness can be and has been extrapolated to human behavior. The experimental psychologist Thomas Bouchard has found a striking behavioral similarity in human identical twins separated early in life and reunited later. Determinism at this level now seems much greater than had been presupposed.[28] It is however very important to us as humans that the degree of chance or indeterminism that we can superpose on a highly deterministic development leaves much room for the "human spirit." The liberty that results is at the heart of research and education. Research has already taught us how to see the other side of the Moon by television; it is now

*The chemical experiments are consistent with a soft determinism. The production of thermal proteins by polymerization of amino acids by heat does not yield a single macromolecular type but, rather, an array of thermal protein molecules of sharply limited heterogeneity. Such a variety admits selection processes.

teaching us to construct colonies in space, and to improve memory in the human brain. This is perhaps enough liberty to satisfy the human spirit. Nonetheless, the material evolutionary embodiment of this liberty has to learn to control its environment so that it can continue to enjoy conditions that will permit it to evolve further, biologically and socially. The test of an adequate environment for that evolution may well be its suitability for all of the species that have emerged along the way and are still here.

CHAPTER 9

This Matter of Mind: Excitability Is Exciting

What are the outstanding scientific problems of the twentieth century? Since we are approaching its end, we can say with considerable confidence that the selection will not be superseded in the remaining years. In physics, the large questions have dealt with quantum theory and relativity; in biology, they have concerned the origin of life and aging. We have learned not only that aging is very much under genetic control but also that one can prolong one's life by adopting appropriate behavior, many aspects of which have been identified.

The intellectual spectrum—including the smallest physical particle, a quantum of energy, and the phenomena of mind—spans all of science. From the perspective of quantum science, the decision-making properties of matter are apparent both at the beginning of evolution and in its latest phase. One of the most exquisite accomplishments of the human mind is its understanding of the nature of matter. As the human mind delves further into matter, it is increasingly finding itself—its own roots.

The modern quantum scientist can rightfully tell us that this is what we should have expected all along. One of them, Per-Olov Löwdin, has suggested that all reality consists of the pairing of sensations and that mapping of the relation between mind and matter is the most fundamental of relations to be studied.[1] The evolution from the elementary particles to the abstraction of mind has been through the material complement of the mind, the brain. Can we look at the mind from the inside by looking at the brain from the inside? The growing recognition of a connected evolutionary sequence increases our optimism that we can do so.

A primary characteristic of mind is the ability to make decisions. This ability, as explained by Koichiro Matsuno, can be traced from the molecules, the smallest particles of matter as we ordinarily know it. The smallest particle of zinc oxide, for example, is a single molecule of this compound. If we break it down further, there results a molecule of metallic zinc and half a gaseous molecule of oxygen of the kind we breathe in air, both radically different from zinc oxide. The smallest particle of table sugar is a molecule of sucrose. If we decompose it—that is, break it down—it becomes something else, like carbon dioxide and water mostly. A teaspoonful of either zinc oxide or sucrose consists of a very large number of these extremely small molecules, perhaps ten to a hundred trillion trillion—or, in mathematical notation, 10^{25} to 10^{26}.

Our concern here is not so much with the number as with the uniqueness of each individual. Every molecule of any one type has its own shape or, in some cases, interchangeable set of shapes. The way a given molecule reacts with another is unique. When it reacts more rapidly with a second molecule than with a third, it is expressing its decision-making ability, as Matsuno phrased it. This molecular selectivity is, when sufficiently traced forward, a root of mind.

The fully evolved mind is the repository of human information. Information at the molecular level has become a popular topic in the last forty years. The business of the mind is to absorb information, to transfer it, to systematize it, and to release it as it is needed. These processes have been studied not just abstractly but also experimentally, at the lower molecular level, in brain research, and in computer science. Biological information exists because of the shapes of molecules and the decision-making phenomena resulting from interactions of those shapes.

The transfer of information through protein vesicles, described in chapter 6, has come to be understood as a "slow" transfer. The faster mode of communication in the brain is through electrical signaling. For this latter kind of process, in its many variations, membranes were originally used and are now needed. The modern cell membrane is composed of proteins and lipids. The lipids permit the making of an efficient barrier, which keeps molecules from diffusing out of cells and slows down the diffusion of unwanted molecules into cells. The cell has evolved to provide that barrier, but it also permits electric current to flow into and out of it, by means of channels of protein, which conduct electric current.

The pure lipid that might originally have been present could have been used for the barrier function of primordial and primitive cell membranes. The lipid could not have served the numerous enzyme functions subsequently needed, could not have conducted electrical signals into and out of the cell, and could not have provided a chemical basis for the individuality of each cell. An unaided purely lipid membrane could not have permitted a variety of small food molecules to diffuse into the interior of the cell from a nutrient exterior.

All of the necessary functions, however, including the barrier function, could have existed in the primitive proteins. This is because protein molecules, and probably primitive ones especially, could have contained chemical groups permitting them to function as lipids as well. It is most likely that the first cells and their membranes were assemblies of proteins—were lipid-like—and that these primitive cells then learned how to make their own, discrete lipid.

This Matter of Mind: Excitability Is Exciting

Membranes for Life and Mind

The cellular unit of the brain, and of the peripheral nervous system, is the excitable cell, which in the brain is the neuron. The emergence of life has been variously attributed primarily to metabolism, membranes, reproduction, or information—or to all of these.*

Membranes are essential structural components of all cells, including the neurons. Biologists who have reasoned that the membrane was a primary requisite for the origin of life are thus indirectly supporting the now experimentally derived suggestion that the origin of life embraces the origin of mind.

That the unit of the brain is a cell that has an electrically active membrane is understood in the same way as the fact that excitable proteinoid microspheres have membranes. The double-layer infrastructure of that membrane has already been shown by the electron microscope (see fig. 4.3).

These membranes have other properties of modern cell membranes. When they are made in salt solution of one concentration and are then transferred to a richer solution, they shrink. When transferred to a more dilute solution, they swell. This mimics osmotic behavior.

*The subject matter to chapter 9 has been worked and reworked for ten to thirty years. Criticism within the laboratory group has been in many ways buttressed by numerous challenges and objections from others. All of this has been helpful.

To recapitulate, the essence of what has been learned has stemmed from moving the questions from their usual context into one of stepwise evolution. The answers include the finding that heated sets of amino acids including dicarboxylic amino acids order themselves into informational proteins and that these polymers avidly assemble themselves into protocells on contact with water. Experiments with laboratory protocells have shown them to be evolvable to modern cells. The protocells have the roots of modern biofunctions including growth, metabolism, reproduction, and membranous electricity.

This last quality and its relative newness sets chapter 9 in the position more of looking forward to new vistas than of discussing old ones. The finding of electrical activity in artificial cells of thermal protein and lecithin was a beginning in experiments repeatedly suggested by the late H. Burr Steinbach, physiologist at the University of Chicago, and by a model for the terrestrial origins of communication (chapter 6).[2]

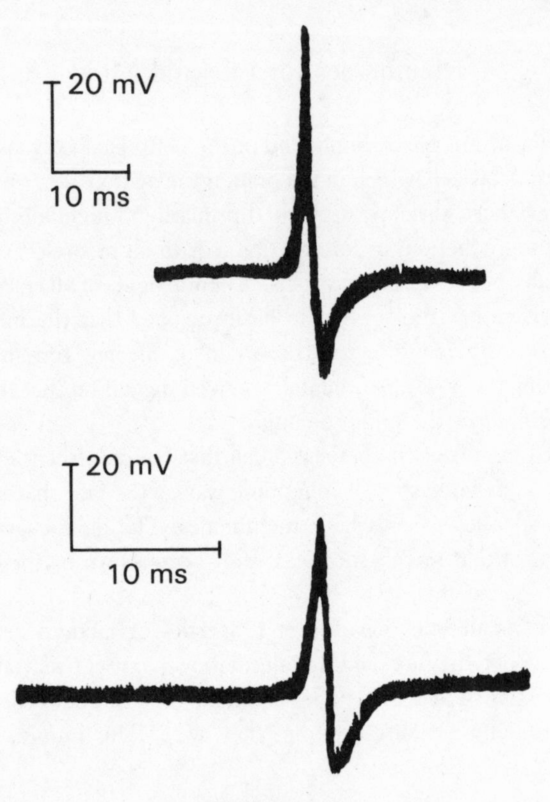

Figure 9.1
Microsphere Spiking

Action potential resembling that of neuron. (Upper) Spiking in crayfish stretch receptor neuron. (Lower) Spiking in microsphere of 2:2:1-proteinoid microsphere. (By Dr. A. Przybylski.)

The electrical behavior of a typical membrane is seen in figures 9.1 and 9.2. When at rest, the membrane displays a potential. Here we see action potentials, also referred to as spikes. At the top of figure 9.1 is an action potential of a crayfish stretch receptor neuron that has been stimulated by electrical current. Immediately below that is one produced by the impalement of a proteinoid microsphere. In figure 9.2, the discharge from a laboratory protocell is above, from a neuron of *Aplysia* sea cucumber, below.

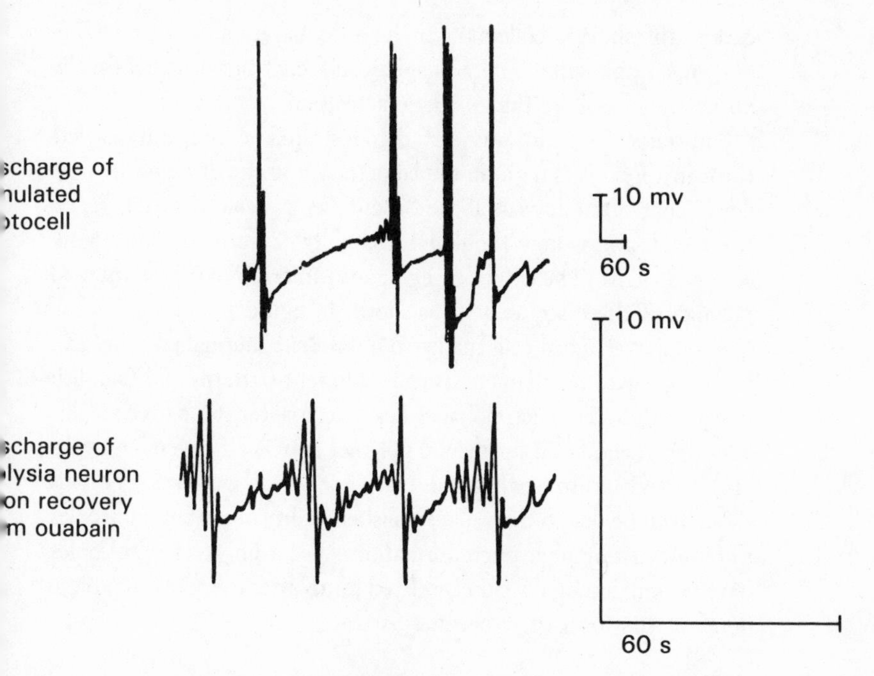

Figure 9.2
Action Potentials

Laboratory protocell above, *Aplysia* neuron below.

The evidence that the boundary of the proteinoid microsphere is a membrane in many ways like that of the modern cell is manifold. It consists of ultrastructures seen in the electron microscope, osmotic behavior, resealability when punctured, selective diffusion of large and small molecules through that boundary, and characteristic electrical behaviors. The microsphere's membrane is thicker than the vast majority of modern membranes, but it is not too thick to display the above properties.

Trains of spikes (action potentials) can be initiated by light, by the kinds of ion shift concentration that occur in living tissue, by chemical action, or by injection of electric current.[3] Electrons, it is now thought, accumulate in recesses in the membrane, itself composed of folded protein molecules; when the number of electrons reaches a certain threshold, a collection of them discharges almost simultaneously in a single spike. Trains of spikes and repetitions thereof usually do not require more than a single activation.

The effect of light was first demonstrated in preparations left through a moonless night in the laboratory into the dawn of the next day. What often appears to be "spontaneous" electrical activity in the membrane is now thought to have its cause in the very weak action of light. The effect of light on a microsphere of a thermal polymer of three amino acids is shown in figure 9.3.

Excitable artificial cells made from different thermal proteins give what appear to be characteristically different patterns. To establish unequivocally that each polymer has its characteristic pattern is difficult, however, because any one polymer may itself engender a variety—though a somewhat limited variety—of patterns. This is thought to be due to the well-established ability of protein molecules to fold into a number of conformations. Accordingly, the molecules have several potentially closely related forms after they are assembled. Some of the forms may become fixed.

Prospects in Research

The proteinoid microspheres can be studied as artificial cells functioning in research to model natural cells. For this purpose, the origin-of-life history may be placed into the background. The resultant understanding of aging, memory, and thought is substantial and intrinsically worthwhile; these processes and the resultant phenomena can be studied in the modern mode, since the retracement experiments are being done here and now.

The prospects for relating the electrical attributes of the brain to molecules were perceived as early as 1960 by Linus Pauling, who noted,

> We may ask what the next step in the search for an understanding of the nature of life will be. I think that it will be the nature of the

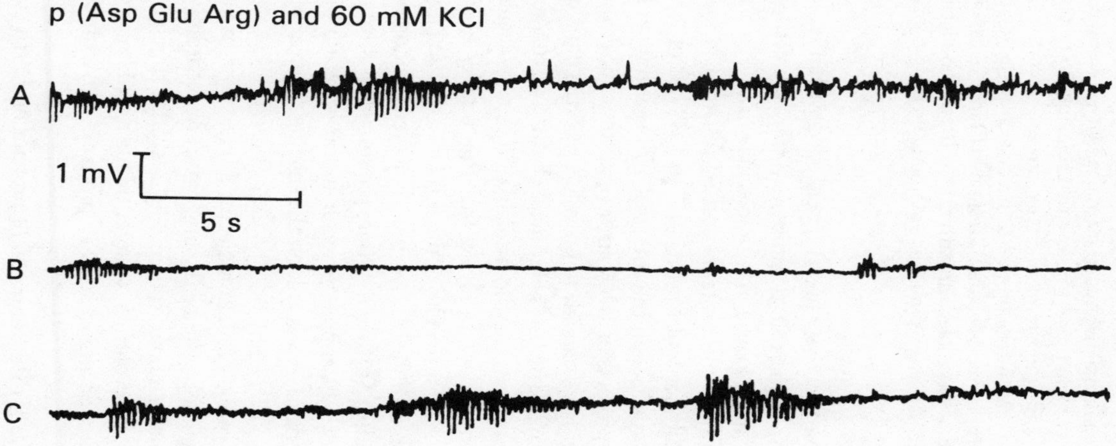

Figure 9.3
Light Activation

Electrical activity begins with illumination (10 lux) at A. It dies off when the light is extinguished (B), and it begins again with reillumination (100 lux) at C.

> electromagnetic phenomena involved in mental activity in relation to the molecular structure of brain tissue. I believe that thinking, both conscious and unconscious, and short-term memory involve electromagnetic phenomena in the brain, interacting with the molecular (material) patterns of the long-term memory, obtained from inheritance or experience. What is the nature of the molecular patterns? What is the mechanism of their interaction? These are problems of structural chemistry that we may now strive to solve.[4]

The idea that mental activity might be encoded within molecules was taken further by Francis O. Schmitt in 1962. Schmitt invoked the variety available in biological macromolecules: "20^{1000} combinations would be possible for proteins and 4^{1000} for RNA—giant polymers could readily account for the 10^{15} to 10^{20} bits of information thought to be processed in the course of a 70-year human life."[5] These numbers for proteins or RNA are supraastronomical; they are obviously more than large enough to provide a material basis for the bits of information in one person's lifetime. Schmitt's calculations are based on the assumption that the available protein molecules are random. That is tantamount to saying that the number of protein molecules available is infinite. The number is finite, but so large that it might as well be infinite. As we saw, however, randomness in proteins is not a reality. At the same time, the storage of information in the brain may not be a simple molecule-for-bit or bit-for-bit relation, but for a materialist this is at least the starting point for investigation.

The number of types of protein molecule did not arise from selection from a random number, since the original array, according to the experiments, was not random. There was instead a much smaller number. During biotic evolution, a larger number could have arisen by the recombination of amino acids in proteins. This number can be estimated.

Realistic estimates for the sum total of proteins on Earth evolving from an initial ten types of amino acids (as suggested by experiments) are in the range of 10^{10} to 10^{30}. Even though minuscule compared with Schmitt's 20^{1000}, these are nonetheless easily large enough to accommodate all the bits of information in a lifetime.

The development of the research laid out by Pauling has been relatively slow. It was first necessary to learn that an extensive, though limited, variety of molecular types are represented by the artificial proteins, that these proteins can assemble into cells, that the cells automatically have membranes, that the membranes display electrical behavior, that the membrane electricity depends upon its protein, and that the electrical patterns may vary with different thermal proteins (the last correlation is yet to be placed on a firm basis).

Aging

Compared with excised cells, the artificial protein cells are remarkably stable and long-lived. That the constituent polymers might confer antiaging properties on proteinoid microspheres, and especially on modern cells, emerged from rather heuristic experiments conducted by Franz Hefti. When Hefti added laboratory polymers to cultures of brain cells of rat embryo, he found two exciting phenomena. The cells lived much longer, and they also stimulated the kind of growth of cellular extensions that characterizes healthy young nerve cells and that connects them.[6]

The polymers that first showed such activity were of a somewhat special kind rich in hydrophobic (water-hating) amino acids. When microspheres were made from such polymers, the resultant units readily projected tubular outgrowths (fig. 9.4). It is for this reason that the polymers were tested on true neurons.

When administered to mice in the Geriatric Research Center of the Veterans Administration Hospital in Sepulveda, California, these same polymers were found greatly to improve memory in these animals. Arthur Cherkin and James Flood used a clever test in which mice were conditioned to avoid a mild foot shock. They were tested for their memory of the buzzer one week after the administration of the shock. In the experiments, various potential memory-enhancing drugs are given at the time the standard foot shock is applied.

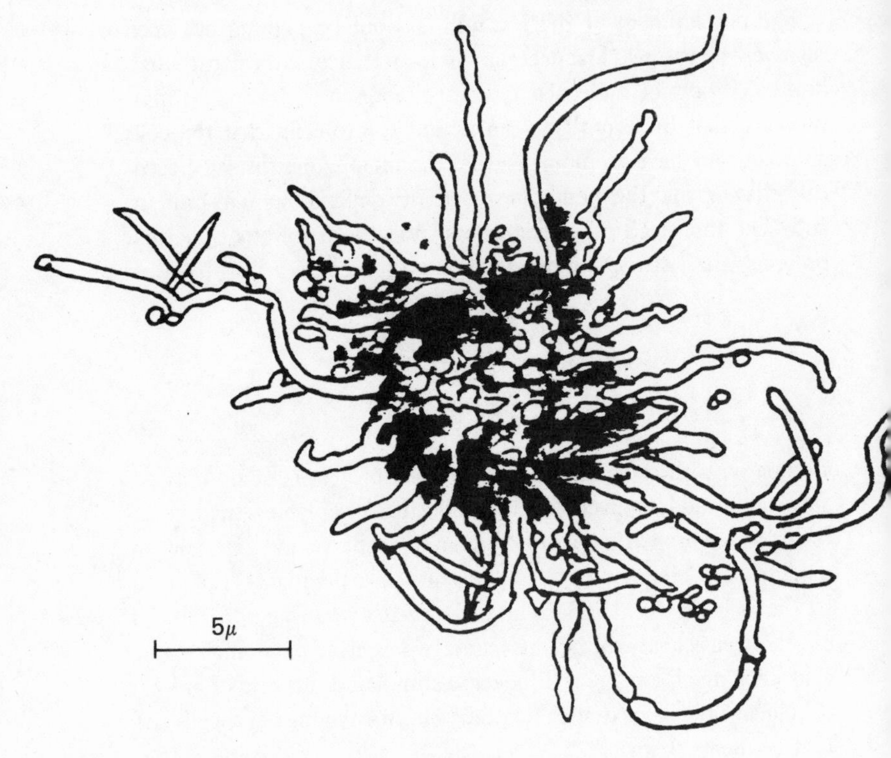

Figure 9.4
Microsphere Outgrowths

A microsphere made from leucine, proline-rich proteinoid projects outgrowths resembling those from neurons.

Although many drugs can enhance or weaken memory to some degree, those that also prolong the life of neurons and stimulate the extension of their outgrowths are of special interest. Understanding the origin of life or the origin of mind, or the origin of life and mind, is not the objective in research like this. However, the possibility of realizing some of man's age-old dreams, such as the prolongation of life and of memory, is coming out of more basic investigation into the chemistry of the first life. An artificial brain, representing a kind of cerebral immortality, has received serious attention, some of it in the scientific literature. The effects of old age appear increasingly to be related to the processes with which the individual animal began its life and with which the evolution of the species began. More and more, we can maintain that the end is influenced by the beginning.

Conversation 3

How would you summarize the main points of what we have been talking about?

Experiments for retracing principal molecular events in prebiotic, protobiotic, and early biotic evolution have indicated a highly nonrandom, albeit not totally determinate, stepwise sequence (as in the flowsheet, fig. 8.1). The resultant laboratory protocells possess properties describing a protometabolism, precisely limited protogrowth, an ability for heterotrophic reproduction, electrical responses to stimuli, and related phenomena. Similar protoorganisms theoretically were subject to further selection to modern organisms by Darwinian evolution. Further experiments partially describe how the necessary ancient protein, already informed by self-ordering of precursor amino acids, underwent complex changes to a genetic coding mechanism.

Fundamental principles identified are (1) attainment of otherwise impossible compounds by steps, (2) self-sequencing of amino acids to informed proteins, and (3) self-organization of the proteins into cells on contact with water.

The experiments reveal that instructions for sequences of amino acids in the first proteins could have come from the amino acids themselves; computerized studies of modern proteins suggest that this process has been inherited. Experiments show that lipid quality in membranes of original cells could have been due to lipoidal side-chains of amino acids.

Conversation 3

How long have you been working in this problem area?

Approximately forty years. Our first experiments on preparing proteins were done in 1950–55; they were essentially a punctuation mark in a small paragraph of evolutionary studies on proteins begun about ten years earlier.

Have you seen much change in the attitudes of those who follow the experimental results?

The change in the receptivity to them has been considerable. When I was first active, in the late 1940s and the 1950s, I was made to feel like a pariah for engaging in this sort of experimental research. The flavor of the responses began to change about 1960–65, I believe after an invited address to the 1959 annual meeting of the American Association for the Advancement of Science and after the appearance of the high school textbook, *Molecules to Man,* in 1964.

The interest in the scientific, or natural, explanation to how life began is spreading vigorously now. Watching what happened has been very educational.

Where does this science go from here?

I see three main vistas. One is synthetic biology. Those students who learn both chemistry and biology and how to employ them integratively will be especially well prepared for a vast new sea of opportunity. This isn't to say that there haven't already been many biologists who know a lot of chemistry and also some chemists who know a lot of biology. There have been, but they haven't, for the most part, used their knowledge in constructionistic investigations.

A second vista is another kind of biotechnology. Proteins are now made by automatic synthesizers and by genetic engineering. The direct production of thermal proteins can also be carried out easily. Each approach has its advantages and disadvantages. A principal advantage of the thermal method is that it permits the expeditious production of a wide variety of proteins that were not made and tested selectively in evolution itself. The production of cells to order is beginning to bring new applications in the microencapsulation industry.

A third vista is synthetic neurobiology.[1] It may lead into anti-aging, memory enhancement, and extracorporeal memory.

NOTES

Preface

1. American Institute of Biological Sciences, *Molecules to Man* (Boston: Houghton Mifflin, 1963) was a book of more than 800 pages, prepared by more than 110 university and high school biology teachers in the Biological Sciences Curriculum Study, which has been the prototype for high school biology textbooks; new editions have appeared since 1963. A recommendable textbook among the very many is Karen Arms and Pamela S. Camp, *Biology* (New York: Holt, Rinehart, and Winston, 1979).

2. Steven Weinberg, *The First Three Minutes* (New York: Basic Books, 1977).

3. Even in their title, *The Left Hand of Creation* (New York: Basic Books, 1983), the authors John D. Barrow and Joseph Silk express this universal bias. On page *x* of their Prologue, they state: "We find only the left-handed amino acids in living things, never their mirror images. This tale of broken symmetries extends from the beginning of time to the here and now."

Conversation 1

1. Alexander I. Oparin, *The Origin of Life on the Earth* (Edinburgh: Oliver and Boyd, 1957) is thorough on the history of the subject.

Chapter 1

1. American Institute of Biological Sciences, *Molecules to Man* (Boston: Houghton Mifflin, 1963).

2. Charles Darwin, *The Origin of Species by Means of Natural Selection, or the Preservation of Favored Races in the Struggle for Life* (New York: Random House, no date), 64. This version of the work includes Darwin's *The Descent of Man* in the same volume; it has appeared in a number of revised versions as well.

3. Darwin considered variation and natural selection of variants. Not much atten-

tion has been paid to the exact nature of variation due to internal factors. One exception was Lancelot L. Whyte, *Internal Factors in Evolution* (New York: George Braziller, 1965).

In his book, Whyte, who was a physicist, states: "I do not understand why no molecular or other biologist has written this book and I have been forced to undertake it" (p. 25). While I was on a visit to Cambridge in 1967, Whyte asked me to meet him in London, which I did, but I was then too immersed in laboratory research to appreciate fully his vision. I would now say that the reason no biologist had written his book is that the education of scientists is (almost necessarily) too fragmented. Physicists know physics and think like physicists, chemists know chemistry and think like chemists, and biologists know biology and think like biologists. Twenty years later, the fragmentation is deeper. In general, internal factors suggesting internal and endogenous control was an unpopular idea in 1965.

4. John Newhouse, *New Yorker Magazine* (10 February 1986):95.

5. Marcel Florkin and Howard S. Mason, *Comparative Biochemistry,* Vols. 1–7 (New York: Academic Press, 1960–64) is a work by leading proponents which develops an emphasis on the unity of biochemistry in exquisite detail.

6. K. Kvenvolden, "Amino and fatty acids in carbonaceous meteorites," in *The Origin of Life & Evolutionary Biochemistry,* ed. K. Dose, S.W. Fox, G.A. Deborin, and T.E. Pavlovskaya (New York: Plenum Press, 1974) is a review of amino acid analyses of meteorites.

7. S.W. Fox, K. Harada, and P.E. Hare, "Amino acid precursors indigenous to lunar samples: accumulated analyses and evaluations of possible contaminations," in *Interactions of Interplanetary Plasma with the Modern and Ancient Moon,* ed. D.R. Criswell and J.W. Freeman (Houston: Lunar Science Institute, 1974) and S.W. Fox, K. Harada, and P.E. Hare, "Accumulated analyses of amino acid precursors in lunar samples," *Geochimica et Cosmochimica Acta 2,* Suppl. 4, 2241–48 contain reviews of amino acid analyses from returned lunar samples.

8. S. Yuasa and J. Oró, "HCN as a possible precursor of the amino acids in lunar samples," in *Science and Scientists,* ed. M. Kageyama, K. Nakamura, T. Oshima, and T. Uchida (Tokyo: Japan Sc. Soc. Press, 1981) contains the following statement by Oró: "It is well known that several amino acids have been found in the acid hydrolyzed hot water extracts of lunar samples (Fox et al., 1972)."

9. Harold Blum, *Time's Arrow and Evolution* (Princeton: Princeton University Press, 1951), 170.

10. R.S. Young, "Prebiological evolution: the constructionist approach to the origin of life," in *Molecular Evolution and Protobiology,* ed. K. Matsuno, K. Dose, K. Harada, and D.L. Rohlfing (New York: Plenum Press, 1984). On p. 48, Dr. Young states:

> Whether or not microspheres truly represent the evolutionary pathway taken in the earliest history of life, they represent a rational approach to the understanding of the crucial events in the origin of life story—the transition of the inanimate to the animate. I would argue that it is only in this fashion that the total story can ever be unraveled. No amount of dissection of contemporary cells will ever unveil this phase of evolution, nor is it likely to be found in the fossil record.

Notes

I have always been puzzled by the fact that of the 100 or so laboratories working in the origin of life field only one has devoted itself in some way to the synthesis of a living system. A few have worked with model systems and none have stated in a grant proposal that their goal was to synthesize life. Perhaps this is a commentary on present-day grant review and the political pressures brought to bear. In any case, if the understanding of the origin of life is a serious subject for scientific study (and surely it is) then the synthesis of a living system, by competent scientific persons and methodology is a necessary subset of the subject.

In my view, the synthesis of a living system is the only hope for development of the complete origin of life story.

In this quotation, Dr. Young is emphasizing that one cannot hope to understand the origin of life by extrapolating backward, through fossils or otherwise.

Chapter 2

1. R.A. Kerr, "Origin of life: new ingredients suggested," *Science* 210 (1980): 42–43.

2. L. Sprague De Camp, *The Great Monkey Trial* (New York: Doubleday, 1968) contains a careful reconstruction of the Scopes trial.

3. Oparin, *The Origin of Life,* 90. Oparin comments: "The resemblance between the objects created by Leduc and living things was no greater than the resemblance between a living person and a marble statue of him. The work emanating from the laboratory of the Mexican investigator A.L. Herrera was of the same nature."

4. A.L. Herrera, "A new theory of the origin of life," *Science* 96, 14 (1942) is a review of Herrera's work which summarizes forty-three years of investigation in a one-page abstract.

5. Albert L. Lehninger, *Biochemistry,* 2nd ed. (New York: Worth Publishers, 1975), 1031.

6. Ibid., 1047.

7. Alexander I. Oparin, *The Origin of Life on the Earth* (Edinburgh: Oliver and Boyd, 1957) contains a discussion of coacervate droplets in chapter 7 and of the action of natural selection to introduce stability into the droplets on pp. 349–57.

8. G. Wald, "The origin of life," *Scientific American* 191, 2 (1954):44–53.

9. S.W. Fox, ed., *The Origin of Prebiological Systems and their Molecular Matrices* (New York: Academic Press, 1965).

10. A.I. Oparin, "Routes for the origin of the first forms of life," *Sub-Cellular Biochemistry* 1 (1971):75–81 reports interesting experiments which describe the synthesis and hydrolysis of starch within the boundaries. This helps support the importance of cellular membranes but, since the enzyme is a modern one, does not help explain "origins."

11. Charles B. Thaxton, Walter L. Bradley, and Roger L. Olsen, *The Mystery of Life's Origin: Reassessing Current Theories* (New York: Philosophical Library, 1984) is a popular scientific book of more than 225 pages which reveals its creationist bias in the Epilogue. A critique of the book appears in S.W. Fox, "Beyond the power of science?," *Quarterly Review Biology* 60 (1985):193–95.

12. Henry M. Morris, *Scientific Creationism* (San Diego: Creation-life Publishers, 1974), 5.

13. M. Ruse, "A philosopher's day in court," in *Science and Creationism*, A. Montagu, ed. (Oxford, Oxford University Press, 1984), 316.

Chapter 3

1. Robert Shapiro, *Origins: A Skeptic's Guide to the Creation of Life on Earth* (New York: Summit Books, 1986). As a biochemically knowledgeable organic chemist, Shapiro was a pioneer in publishing a nontechnical book that explains why proteins probably arose first. On p. 282, he states: "Protein begat RNA, and then both begat DNA."

2. Francis Crick, *Life Itself: Its Origin and Nature* (New York: Simon and Schuster, 1981).

3. For the specific finding of the difference in sickle-cell hemoglobin, see V.M. Ingram, "Genetic mutations in human hemoglobin: the chemical difference between normal and sickle-cell hemoglobin," *Nature* 180 (1957):326. For a broader treatment, see Vernon M. Ingram, *The Hemoglobins in Genetics and Evolution* (New York: Columbia University Press, 1963). The modern study of amino acid sequence in proteins was punctuated from the author's laboratory by a prospectus in S.W. Fox, "Terminal amino acids in peptides and proteins," *Advances Protein Chemistry* 2 (1945):155–77. From this followed structural analyses of insulin, hemoglobin, and other proteins, according to J. Rosmus and Z. Deyl, "The methods for identification of N–terminal amino acids in peptides and proteins. Part B," *Chromatography Reviews* 13 (1972):221–339.

4. G. Ungar, ed., *Molecular Mechanisms in Memory and Learning* (New York: Plenum Press, 1970), viii. See also: D. Wilson, "Scotophobin resurrected as a natural peptide," *Nature* 320 (1986):312–14.

5. The work of Schiff, Schaal, and others is reviewed by E. Katchalski, *Advances Protein Chemistry* 6 (1951):123–85. Katchalski, a highly regarded polymer chemist, was the fourth president of Israel under the name of Ephraim Katzir, and director in the Weizmann Institute in Rehovot.

6. J. Kovacs, H. Kovacs, F. Koenyves, J. Csaszar, T. Vajda, and H. Mix, "Chemical studies of polyaspartic acids," *Journal Organic Chemistry* 26 (1961):1084–91.

7. A. Vegotsky, K. Harada, and S.W. Fox, "The characterization of polyaspartic acid and some related compounds," *Journal American Chemical Society* 80 (1958):3361–66.

8. S.W. Fox and K. Harada, "Thermal copolymerization of amino acids to a product resembling protein," *Science* 128 (1958):1214.

9. Alexander I. Oparin, *The Origin of Life on the Earth* (Edinburgh: Oliver and Boyd, 1957), 289–90. Oparin was a leader in the inference that proteins arose first, but he did not entertain the possibility, shown by experiments, that the first proteins were orderly, i.e., informed. That suggestion came from experiments that were reported from 1958 on.

10. S.W. Fox and H. Wax, "Enzymic synthesis of peptide bonds. III. The relative effects of some amino acids and their acyl substituents," *Journal American Chemical Society* 72 (1950):5087–90; S.W. Fox, M. Winitz, and C.W. Pettinga, "Enzymic synthesis of peptide bonds. VI. The influence of residue type on papain-catalyzed reactions of some benzoylamino acids with some amino acid anilides," *Journal American Chemical Society* 75 (1953):5539–42.

11. One of the first theorists, after Wald, to reason proteins first was another Nobel laureate, Fritz Lipmann. See S.W. Fox, ed., *The Origins of Prebiological Systems and of their Molecular Matrices* (New York: Academic Press, 1965), 259–73.

12. One early review of the self-ordering of amino acids is by J. Hartmann, M.C. Brand, and K. Dose, "Formation of specific amino acid sequences during thermal polymerization of amino acids," *BioSystems* 13 (1981):141–47.

13. Another question that sometimes concerns specialists is the origin of chirality (optical activity). The many answers that are at hand (Sidney W. Fox and Klaus Dose, *Molecular Evolution and the Origin of Life* [New York: Marcel Dekker, 1977], 268–75) include some that span prebiotic amino acids (K. Harada, "Origin and development of optical activity of organic compounds on the primordial Earth," *Naturwissenschaften* 57 [1970]: 114–19) and later stages of biological evolution (Werner Langenbeck, *Die Organische Katalysatoren* [Berlin: Springer, 1935] and S. Fox and K. Dose, *Molecular Evolution,* 272). It is however true that the biofunctions to be described in chapter 6 have been identified in microspheres composed of thermal proteins in turn composed mainly of racemic amino acids. Optical activity was shown, as no other kind of study could show, not to be necessary at the early cellular stages.

Enzyme-like properties are described in K. Dose, "Self-instructed condensation of amino acids and the origin of biological information," *International Journal of Quantum Chemistry, Quantum Biology Symposium* 11 (1984):91–101.

Chapter 4

1. D. Nachmansohn, "Chemical factors controlling nerve activity," *Science* 134 (1961):1962–68 and D. Nachmansohn, "Proteins in excitable membranes," *Science* 168 (1970):1059–66 were two papers ahead of their time.

2. F. O. Schmitt, "Macromolecular interaction patterns in biological systems," *Proceedings American Philosophical Society* 100 (1956):476–86.

3. Proteinoid microspheres stable over a range of pH were first described in S.W.

Fox and S. Yuyama, "Effects of the Gram stain on microspheres from thermal polyamino acids," *Journal Bacteriology* 85 (1963):279–83.

4. E. Tobach and T.C. Schneirla, "The biopsychology of social behavior of animals," in *Biological Basis of Pediatric Practice,* ed. R.E. Cooke and S. Leven (New York: McGraw-Hill, 1968), 68.

Chapter 5

1. The model of primordial mating of proteinoid microspheres is described in S.W. Fox, L.L. Hsu, S. Brooke, T. Nakashima, and J.C. Lacey, Jr., "Experimental models of communication at the molecular and microsystemic levels," *International Journal Neuroscience* 3 (1972):183–92.

2. L.L. Hsu, S. Brooke, and S.W. Fox, "Conjugation of proteinoid microspheres," *Currents Modern Biology* 4 (1971):12–25.

3. Ibid.

4. A first report on proliferation (protoreproduction) of proteinoid microspheres is in S.W. Fox, R. J. McCauley, and A. Wood, "A model of primitive heterotrophic proliferation," *Comparative Biochemistry and Physiology* 20 (1967):773–78.

5. The sequence showing motility has appeared in a number of reviews. One is S.W. Fox, "The origins of behavior in macromolecules and protocells," *Comparative Biochemistry and Physiology* 67B (1980):423–36.

6. The association of enzyme-like activities is described in technical detail in S.W. Fox, "Metabolic microspheres," *Naturwissenschaften* 67 (1980):378–93, and in T. Nakashima, "Metabolism of proteinoid microspheres," *Topics Current Chemistry* 139 (1987):57–81.

7. The kind of particle that has the activity to synthesize the bonds of protein and of nucleic acids was first made in 1962 by Yuyama (see Note 3, Chapter 4, p.173) and later in greater profusion by Waehneldt and Fox as reported in T.V. Waehneldt and S.W. Fox, "The binding of basic proteinoids with organismic or thermally synthesized polynucleotides," *Biochimica et Biophysica Acta* 160 (1968):239–45. Synthetic ability is described in S.W. Fox, J.R. Jungck, and T. Nakashima, "From protein microsphere to contemporary cell formation of internucleotide and peptide bonds by proteinoid particles," *Origins of Life* 5 (1974):227–37.

Chapter 6

1. J.J. Copeland, "Yellowstone thermal myxophyceae," *Annals New York Academy of Sciences* 36 (1936):1–232.

2. A principal modern description of the situation for hot springs beneath the ocean is found in J.B. Corliss, J.A. Baross, and S.E. Hoffman, "An hypothesis concerning the relationship between submarine hot springs and the origin of life on Earth,"

Oceanologica Acta N°SP (1981):59–69. This kind of locale was earlier suggested, by laboratory experiments, in S.W. Fox, "The chemical problem of spontaneous generation," *Journal Chemical Education* 34 (1957):472–79.

3. Copeland, "Yellowstone thermal myxophyceae."

4. E.S. Barghoorn and S.A. Tyler, "Fossil organisms from Precambrian sediments," *Annals New York Academy of Sciences* 108 (1963):451–52.

5. Comparison of ancient microfossils and microspheres can be seen in S.W. Fox and Klaus Dose, *Molecular Evolution and the Origin of Life,* rev. ed. (New York: Marcel Dekker, 1977), 304 or in S.W. Fox, "Proteinoid experiments and evolutionary theory," in *Beyond Neo-Darwinism, An Introduction to the New Evolutionary Paradigm,* ed. M. W. Ho and P. T. Saunders (London: Academic Press, 1984), 33–35.

6. Description of the artificially fossilized microspheres compared to similarly treated organisms is in S. Francis, L. Margulis, and E.S. Barghoorn, "On the experimental silicification of microorganisms. II. On the time of appearance of eukaryotic organisms in the fossil record," *Precambrian Research* 6 (1978):65–100.

7. Arthur Kornberg, *DNA Replication* (San Francisco: W. H. Freeman Co., 1980), 87–88.

8. T.R. Cech, "RNA splicing: three themes with variations," *Cell* 34 (1983):713–16. The unanswered questions for evolution from an RNA beginning have been dramatized by the following comparison chart:

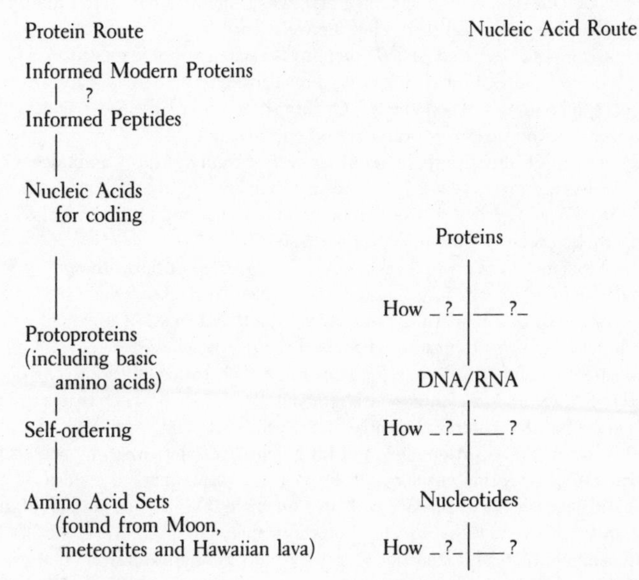

9. Gilbert Ling, *In Search of the Physical Basis of Life* (New York: Plenum Press, 1984), 498. In this book, Ling discusses the models favoring lipids or proteins as the seats of electrical potential in cell membranes. From experiments with the microsphere model, he states: "There is now no doubt that the primary seats of the generation of the potentials are proteins, not phospholipids" (p. 498).

10. D. Nachmansohn, "Proteins in excitable membranes," *Science* 168 (1970): 1059–65.

Chapter 7

1. P.O. Löwdin, ed., *International Journal of Quantum Chemistry, Quantum Biology Symposium* 11 (1984), contains, in pp. 1–135, a report of a conference on "The Origin of Life and Mind," and, in pp. 31–43, an explanation of the special relationship of mind to quantum science.

2. Francis Crick, *Life Itself: Its Origin and Nature* (New York: Simon and Schuster, 1981).

3. Alexander I. Oparin, *The Origin of Life on the Earth,* (Edinburgh: Oliver and Boyd, 1957), 68–69.

4. Ibid., 69.

5. Crick, *Life Itself,* 148. Crick states: "There seem to me to be two different criticisms of Directed Panspermia. The first, which my wife has voiced more than once, is that it is not a real theory but merely science fiction."

6. The computerized study of 2898 proteins by their amino acid sequences is found in O.C. Ivanov and B. Förtsch, "Universal regularities in protein primary structure: preference in bonding and periodicity," *Origins of Life* 17 (1986):35–49. Professor G. Braunitzer is acknowledged for support and encouragement.

7. A. Graham Cairns-Smith, *Seven Clues to the Origin of Life* (Cambridge: Cambridge University Press, 1985), 116. This is a later book by the author.

8. S.W. Fox, "Thermal polymerization of amino acids and production of formed microparticles on lava," *Nature* 201 (1964):336–37.

9. F. Lipmann, "Projecting backward on the evolution of biosynthesis," in S.W. Fox, ed., *The Origins of Prebiological Systems* (New York: Academic Press, 1965).

One of the earliest biochemists to question that DNA or RNA preceded protein in evolution was Fritz Lipmann, whose Nobel Prize was awarded for his brilliant investigations on protein synthesis. Lipmann states: "My basic motivation for entering into this discussion is an uneasy feeling about the apparent tenet that a genetic information transfer system is essential at the very start of life" (p. 259).

He then mentions: "Strominger and his group have shown that, by consecutive addition through specific enzymes, up to six amino acids may be condensed to a polypeptide in a specified sequence without a template. This is some kind of information transfer in which RNA does not enter into the picture" (p. 271).

Lipmann, and later his student Horst Kleinkauf, showed extensively that precise polypeptide synthesis could occur without any direct contact with RNA. The synthesis, we now see, was controlled endogenously by the reactants and one or more enzymes. Also, Lipmann explained in his summary that the backward projection mentioned in his title would have made "it necessary to make assumptions which seem difficult or perhaps impossible to verify" (pp. 272–73). He chose rather to recast the evolution of protein in a forward direction.

10. Various overviews related more or less closely to the proteinoid theme in this volume have been published. Although each of these authors has been close to the main study, each has developed a broad, independent view. Prominent ones are:

K. Dose, "Molecular evolution and protobiology: an overview," in *Molecular Evolution and Protobiology,* K. Matsuno, K. Dose, K. Harada, and D.L. Rohlfing, eds. (New York: Plenum Press, 1984), 1–9. This review lists unanswered questions. Also: J. Hartmann, M.C. Brand, and K. Dose, "Formation of specific amino acid sequences during thermal polymerization of amino acids," *BioSystems* 13 (1981):141–47.

J.L. Fox, "Origins," in *Molecular Evolution: Prebiological and Biological,* D.L. Rohlfing and A.I. Oparin, eds. (New York: Plenum Press, 1972), 23–34; J.L. Fox, "Evolution and enzymes," in *Molecular Evolution and Protobiology,* K. Matsuno, K. Dose, K. Harada, and D.L. Rohlfing, eds. (New York: Plenum Press, 1984), 331–36.

R.F. Fox, "A non-equilibrium thermodynamical analysis of the origin of life," in *Molecular Evolution, Prebiological and Biological,* D.L. Rohlfing, and A.I. Oparin, eds. (New York: Plenum Press, 1972), 79–99; R.F. Fox, *Biological Energy Transduction: The Uroboros,* (New York: John Wiley and Sons, 1982); R.F. Fox, "The Uroboros" in *Molecular Evolution and Protobiology,* K. Matsuno, K. Dose, K. Harada, and D.L. Rohlfing, eds. (New York: Plenum Press, 1984), 413–20.

T.O. Fox, "Evolution of levels of evolution," in *Molecular Evolution: Prebiological and Biological,* D.L. Rohlfing, and A.I. Oparin, eds. (New York: Plenum Press, 1972), 35–42; T.O. Fox, "Evolving evolution," in *Molecular Evolution and Protobiology,* K. Matsuno, K. Dose, K. Harada, and D.L. Rohlfing, eds. (New York: Plenum Press, 1984), 387–96.

K. Harada, *Kagaku Shinka* [in Japanese, Chemical Evolution-Chemical Basis of the Origin of Life] (Tokyo: Kyoritsu Publishing Co., 1971); K. Harada, "Some early historical aspects of the thermal polycondensation of amino acids," in *Molecular Evolution and Protobiology,* K. Matsuno, K. Dose, K. Harada, and D.L. Rohlfing, eds. (New York: Plenum Press, 1984), 1–28.

L.L. Hsu, "Conjugation of proteinoid microspheres: a model of primordial recombination," in *Molecular Evolution: Prebiological and Biological,* D.L. Rohlfing, and A.I. Oparin, eds. (New York: Plenum Press, 1972), 371–78; L.L. Hsu, "Concepts of prebiological evolution: their implications on natural selection and time course of evolution," in *Molecular Evolution and Protobiology,* K. Matsuno, K. Dose, K. Harada, and D.L. Rohlfing, eds. (New York: Plenum Press, 1984), 397–412.

J.R. Jungck, "The adaptationist programme in molecular evolution: the origins of genetic codes," in *Molecular Evolution and Protobiology,* K. Matsuno, K. Dose, K. Harada, and D.L. Rohlfing, eds. (New York: Plenum Press, 1984), 345–64.

J.C. Lacey, Jr., and D.W. Mullins, Jr., "Proteins and nucleic acids in prebiotic evolution," in *Molecular Evolution: Prebiological and Biological,* D.L. Rohlfing and A.I. Oparin, eds. (New York: Plenum Press, 1972), 171–88. This paper uses thorough chemical reasoning on why proteins emerged before nucleic acids, the same theme as in Shapiro's 1986 book (p. 51 in this book). J.C. Lacey, Jr., and D.W. Mullins, Jr., "The genetic anticode: the role of thermal proteinoids in development of an hypothesis," in *Molecular Evolution and Protobiology,* K. Matsuno, K. Dose, K. Harada, and D.L. Rohlfing, eds. (New York: Plenum Press, 1984), 267–82.

K. Matsuno, "Protobiology: a theoretical synthesis," in *Molecular Evolution and Protobiology,* K. Matsuno, K. Dose, K. Harada, and D.L. Rohlfing, eds. (New York: Plenum Press, 1984), 433–64. This author-editor has published numerous papers on protobiology from the view of a theoretical physicist.

T. Nakashima, "Protoribosomes," in *Molecular Evolution and Protobiology,* K. Matsuno, K. Dose, K. Harada, and D.L. Rohlfing, eds. (New York: Plenum Press, 1984), 215–31; T. Nakashima, "Metabolism of proteinoid microspheres," *Topics Current Chemistry* 139 (1987):57–81.

D.L. Rohlfing, "The development of the proteinoid model for the origin of life," in *Molecular Evolution and Protobiology,* K. Matsuno, K. Dose, K. Harada, and D.L. Rohlfing, eds. (New York: Plenum Press, 1984), 29–43. An insightful analysis of thinking, pro and con.

11. Papers by Eigen include M. Eigen, "Selforganization of matter and the evolution of biological macromolecules," *Naturwissenschaften* 58 (1971):465–523 and M. Eigen, W. Gardiner, P. Schuster, and R. Winkler-Oswatitsch, "The origin of genetic information," *Scientific American* 244, 4 (1981):88–118.

12. Robert Shapiro, *Origins: A Skeptic's Guide to the Creation of Life on Earth* (New York: Summit Books, 1986), 165–66.

13. Manfred Eigen, "The physics of molecular evolution," in *The Molecular Evolution of Life,* H. Baltscheffsky, H. Jörnvall, and Rigler, eds. (Cambridge: Cambridge University Press, 1986), 25.

14. For orientation in Orgel's approach, the following are recommended: T. Inoue and L.E. Orgel, "A nonenzymatic RNA polymerase model," *Science* 219 (1983):859–862 and L.E. Orgel and R. Lohrmann, "Prebiotic chemistry and nucleic acid replication," *Accounts of Chemical Research* 7 (1974):368–77.

15. G. Wald, "The origin of life," *Scientific American* 191, 2 (1954): 51.

16. Shapiro, *Origins,* 281.

17. Ibid., 282.

18. Ibid., 313.

19. Anonymous, *New Yorker Magazine* (17 February 1986): 105.

20. Personal communication.

21. O.C. Ivanov and B. Förtsch, "Universal regularities in protein primary structure: preference in bonding and periodicity," *Origins of Life* 17 (1986): 35–49.

22. The possibility of collecting objections to the proteinoid theory into a single chapter was considered until it became clear that such objections are taken up throughout the presentations. One exception to this is the criticism in P.A. Temussi, L. Paolillo, L. Ferrara, E. Benedetti, and S. Andini, "Structural characterization of thermal prebiotic polypeptides," *Journal Molecular Evolution* 7 (1976):105–10. The

answer to Temussi is referred to in both J. Hartmann, M.C. Brand, and K. Dose, "Formation of specific amino acid sequences during thermal polymerization of amino acids," *BioSystems* 13 (1981):141–47 and in S.W. Fox, "Response to comments on thermal polypeptides by P.A. Temussi et al.," *Journal Molecular Evolution* 8 (1976):301–304. The essence of the answer is that a modern cell must have *evolved* from a protocell.

For comments like those of the Temussi kind, we are dealing not so much with specific criticisms as we are dealing with alternative paradigms. The principal paradigms are three:

1. Biblical (King James)
2. Indeterminate emergence and evolution
    a. Nucleic acids first
    b. Cells-last
    c. Randomness
3. Determinate emergence and evolution
    a. Proteins first
    b. Cells-early
    c. Nonrandomness

Chapter 8

1. Charles Darwin, *The Origin of Species by Means of Natural Selection, or the Preservation of Favored Races in the Struggle for Life* (New York: Random House, no date), 372.

2. Jacques Monod, *Chance and Necessity,* trans. A. Wainhouse (New York: Alfred A. Knopf, 1971), 98. Monod's book appeared first in French in 1969.

3. M. Eigen, "Selforganization of matter and the evolution of biological macromolecules," *Naturwissenschaften* 58 (1971): 465–523.

4. Francis Crick, *Life Itself: Its Origin and Nature* (New York: Simon and Schuster, 1981), 81.

5. Ibid., 83.

6. Alexander I. Oparin, *The Origin of Life on the Earth* (Edinburgh: Oliver and Boyd, 1957), 289–90. That the first proteins or polypeptides were orderly was crucial to the emergence of living systems, and recognition of the possibility was crucial to understanding the emergence of living organisms. Or, in other words, the evolution on which Oparin built his argument had, as the experiments show, to precede organisms, not be prolonged after the emergence.

7. C.H. Waddington, "The principle of archetypes in evolution," in *Mathematical Challenges to the Neo-Darwinian Interpretation of Evolution,* eds. P.S. Moorhead and M.M. Kaplan (Philadelphia: Wistar Institute Press, 1967), 113–15.

8. Salvador E. Luria, *Life, The Unfinished Experiment* (New York: Charles Scribner's Sons, 1973), 20, 117.

9. M. Eden, "Inadequacies of neo-Darwinian evolution as a scientific theory," in *Mathematical Challenges to the Neo-Darwinian Interpretation of Evolution,* eds. P.S. Moorhead and M.M. Kaplan (Philadelphia: Wistar Institute Press, 1967), 5–12.

10. A. Szent-Györgi, "The evolutionary paradox and biological stability," in *Molecular Evolution: Prebiological and Biological,* eds. D.L. Rohlfing and A.I. Oparin (New York: Plenum Press, 1972), 111–12.

11. Ibid.

12. Mae-Wan Ho and Peter T. Saunders, eds., *Beyond Neo-Darwinism, an Introduction to the New Evolutionary Paradigm* (London: Academic Press, 1984), 5. The volume contains fourteen articles by fourteen authors. The quotation is from the introductory chapter, "Pluralism and convergence in evolutionary theory," by Mae-Wan Ho and Peter T. Saunders.

13. R.T. Bakker, "Evolution by revolution," *Science* 85 (1985):78.

14. Gordon Rattray Taylor, *The Great Evolution Mystery* (New York: Harper and Row, 1983). Taylor states: "If the basic assumption that evolutionary variety depends on chance arrangements of such entities (polynucleotides) is proved false then the lynch pin of evolutionary theory as it has been known in this century is gone" (p. 174). Later, he adds: "Darwinism is not so much a theory as a sub-section of some theory as yet unformulated" (p. 233). These two quotations are of course related; it was necessary to recognize the results of experiments that showed that chance arrangements of nucleotides are not the motor of evolutionary change. As we have seen, amino acids were involved rather than nucleotides, and those results were not chance (random) arrangements.

15. Ernst Mayr, "Prologue. Some thoughts on the history of the evolutionary synthesis," in *The Evolutionary Synthesis,* eds. E. Mayr and W.B. Provine (Cambridge: Harvard University Press, 1980). Mayr states in a footnote: "The term 'Darwinism' in the following discussions refers to the theory that selection is the only direction-giving factor in evolution" (p. 3).

16. Ernst Mayr, *The Growth of Biological Thought: Diversity, Evolution, and Inheritance* (Cambridge: Harvard University Press, Belknap Press, 1982), 567. Following a discussion of Julian Huxley's *evolutionary synthesis* and an enlarged view of evolutionary variety, Mayr talks of the "decline of the concept of 'mutation pressure' and its replacement by a heightened confidence in the powers of natural selection, combined with a new realization of the immensity of genetic variation in natural populations." While this perception was a factual one, it signaled a further departure from Darwin's reliance on "Natural Selection" as the "most important, but not the exclusive, means of modification" and a further departure from grasping Darwin's "causes and laws of variation," or rather the internal directedness of internally limited boundaries.

17. Ho and Saunders, eds., *Beyond Neo-Darwinism,* 6. Ho and Saunders quote from T.H. Morgan, *What is Darwinism?* (New York: W.W. Norton, 1929), 77. Theirs is a nested quotation since Ho and Saunders couch Morgan's comments in significant relationships. Morgan's statement is:

> When, if ever, the whole story can be told, the problem of adaptation of the organism to its environment and the coordination of its parts may appear to be a self-contained progressive elaboration of chemical com-

pounds—a process no more fortuitous than the constitution of the earth or its revolution about the sun. The outcome would be as determined as any natural event, subject always to the principle of survival.

18. A. Montagu, ed., *Science and Creationism* (Oxford: Oxford University Press, 1984), 365–92. The Decision states:

> The emphasis on origins as an aspect of the theory of evolution is peculiar to creationist literature. Although the subject of origins of life is within the province of biology, the scientific community does not consider origins of life a part of evolutionary theory. The theory of evolution assumes the existence of life and is directed to an explanation of *how* life evolved. Evolution does not presuppose the absence of a creator or God and the plain inference conveyed by section 4 is erroneous.[23]

Note 23 states that:

> The idea that belief in a creator and acceptance of the scientific theory of evolution are mutually exclusive is a false premise and offensive to the religious views of many. (Hicks) Dr. Francisco Ayala, a geneticist of considerable renown and a former Catholic priest who has the equivalent of a Ph.D. in theology, pointed out that many working scientists who subscribed to the theory of evolution are devoutly religious.

This note helps to explain why evolutionary theory and "origins" theory have indeed been separate. It raises for some viewers the question of a continuing failure to separate religion and the state—here solidified on advice into a Federal legal opinion.

The evolutionary principles alluded to have been discussed as "Protobiological Self-Organization" in Rome in 1984, and published in the book *Structure in Motion: Membranes, Nucleic Acids & Proteins,* ed. E. Clementi, G. Corongiu, M.H. Sarma, and R.H. Sarma (Guilderland, N.Y.: Adenine Press, 1985) and again later in S.W. Fox, "The evolutionary sequence: origin and emergences," *American Biology Teacher* 48, 3 (1986):140–49, 169. In the latter publication, explanation of this and related matters is depicted as proceeding from Dr. Clementi to Pope John Paul II in a special papal audience. Dr. Ayala was not present as part of this scientific community. The principles of stepwiseness, self-ordering, and selforganization appear to be repeated in each living generation.

19. O.C. Ivanov and B. Förtsch, "Universal regularities in protein primary structure: preference in bonding and periodicity," *Origins of Life* 17 (1986): 47.

20. Thomas S. Kuhn, "Newton's optical papers," in *Isaac Newton's Papers & Letters on Natural Philosophy and Related Documents* (Cambridge: Harvard University Press, 1978), 27–45.

21. William Herbert Carruth, "Each in His Own Tongue," in *One Hundred and One Famous Poems* (Chicago: R.J. Cook, 1923), 104.

22. Albert Einstein, quoted in R.W. Clark, *Einstein, the Life and Times* (New York: The World Publishing Co., 1971), 340.

23. W. Heisenberg in Max Born, *The Born-Einstein Letters* (New York: Walker and Co., 1971), *x*.

24. Ibid., 148–49.

25. Ronald W. Clark, *Einstein, The Life and Times* (New York: The World Publishing Co., 1971), 346–47.

26. Joseph Needham, *Beyond Neo-Darwinism* (Cambridge: Cambridge University Press, 1984), *viii.*

27. Commentary to Clifford M. Will, *Was Einstein Right?* (New York: Basic Books, 1986).

28. A thoroughly reasoned analysis of Chance and Necessity is found in Linus Pauling and E. Zuckerkandl, "Chance in evolution—some philosophical remarks," in *Molecular Evolution: Prebiological and Biological,* D.L. Rohlfing and Alexander I. Oparin, eds. (New York: Plenum Press, 1972), 113–26.

Chapter 9

1. Per-Olov Löwdin, "Some aspects of quantum theory, consciousness of the mind and free will," in *Selforganization,* Sidney W. Fox, ed. (Guilderland: Adenine Press, 1986), 13–31 is a paper in which Löwdin discusses physical reality as dealing with the pairing of sensations.

2. Y. Ishima, A.T. Przybylski, and S.W. Fox, "Electrical membrane phenomena in spherules from proteinoid and lecithin," *BioSystems* 13 (1981):243–51.

In further study, other properties were observed and the lecithin was found often to be superfluous. A review is in A.T. Przybylski and Sidney W. Fox, "Electrical phenomena in proteinoid cells," in *Modern Bioelectrochemistry,* F. Gutmann and H. Keyzer, eds. (New York: Plenum Press, 1986), 377–96.

Since the cells are artificial their constitution is controllable. They promise the possibility of standardizing tests of relationship between structure and behavior. See G. Vaughan, A.T. Przybylski, and S.W. Fox, "Thermal proteinoids as excitability-inducing materials," *BioSystems* 20 (1987):219–23.

3. A.T. Przybylski, "Excitable cell made of thermal proteinoid," *BioSystems* 17 (1986):281–88.

4. Linus Pauling, *The Nature of the Chemical Bond,* 3rd ed. (Ithaca: Cornell University Press, 1960), 570.

5. F.O. Schmitt, "Psychophysics considered at the molecular and submolecular levels," in *Horizons in Biochemistry: Albert Szent-Györgyi Dedicatory Volume,* M. Kasha and B. Pullman, eds. (New York: Academic Press, 1962), 437–57.

6. F. Hefti, J. Hartikka, E. Junard, P. Strang, A. Przybylski, G. Vaughan, and S.W. Fox, "Neurotrophic actions of thermal (artificial) proteins," *Society for Neuroscience* (1987 Annual Meeting Abstract):443.1.

Conversation 3

1. The beginning visualization of this goal is a quarter-century old. It was perhaps most clearly expressed by a famous British neuroanatomist: John Zachary Young, *A*

Model of the Brain (Oxford: Oxford Clarendon Press, 1964). In an introductory section entitled "Making and remaking nervous systems," Young stated:

> We are certainly a long way from being able to make nervous systems to order, but we begin to see, in outline, the principles of how to do so.
>
> I cannot pretend to say how this can be done, or that it will be done. The point is that if we believe that our analysis of living systems is on the right lines, there is surely no reason why we should not proceed with the logical next step of constructing similar systems. This, incidentally, applies also to the study of the origin and production of living matter itself.
>
> The whole discussion takes a new turn, however, when we consider that in such attempts at synthesis there is no need simply to follow nature's way exactly, or to use the same materials, which take millions of years to fashion by natural selection. Man's way is to find other materials and by short cuts produce what he calls "machines" that do the work more easily for him.

Young was in effect saying that if life and mind are not miracles, they were constructed by the blind processes of evolution, without any goal. If evolution could do it that way man can also do it much faster and probably better. Since artificial cells having some of the properties of neurons can be constructed in steps, and since some of the principles of selforganization for both natural and artificial cells have been charted, the road of investigation is increasingly open even in this century.

INDEX

Index

Index

Index

Index